CAMBRIDGE LIBRARY COLLECTION

Books of enduring scholarly value

Earth Sciences

In the nineteenth century, geology emerged as a distinct academic discipline. It pointed the way towards the theory of evolution, as scientists including Gideon Mantell, Adam Sedgwick, Charles Lyell and Roderick Murchison began to use the evidence of minerals, rock formations and fossils to demonstrate that the earth was older by millions of years than the conventional, Bible-based wisdom had supposed. They argued convincingly that the climate, flora and fauna of the distant past could be deduced from geological evidence. Volcanic activity, the formation of mountains, and the action of glaciers and rivers, tides and ocean currents also became better understood. This series includes landmark publications by pioneers of the modern earth sciences, who advanced the scientific understanding of our planet and the processes by which it is constantly re-shaped.

The Tower of Pelée

Born in Hungary, the geologist Angelo Heilprin (1853–1907) moved with his family to the United States as a boy. He later left New York to study natural sciences in distinguished European institutions, and went on to hold academic positions in Philadelphia and ultimately at Yale. His teaching duties were interspersed with expeditions to Yucatan, Greenland and other places of geological interest. This 1904 study, complemented by unique photographs, depicts his third visit to the island of Martinique in the aftermath of the devastating eruptions of Mount Pelée in 1902. Heilprin documents the temporary formation of Pelée's 'tower', a monolith of lava that grew rapidly after the eruptions, reaching a height of 300 metres before its collapse. Heilprin also summarises the chief features of volcanic eruptions and directs readers to his 1903 publication *Mont Pelée and the Tragedy of Martinique* (also reissued in this series) for further information.

The Tower of Pelée

New Studies of the Great Volcano of Martinique

Angelo Heilprin

CAMBRIDGE
UNIVERSITY PRESS

CAMBRIDGE
UNIVERSITY PRESS

University Printing House, Cambridge, CB2 8BS, United Kingdom

Cambridge University Press is part of the University of Cambridge.
It furthers the University's mission by disseminating knowledge in the pursuit of
education, learning and research at the highest international levels of excellence.

www.cambridge.org
Information on this title: www.cambridge.org/9781108082815

© in this compilation Cambridge University Press 2015

This edition first published 1904
This digitally printed version 2015

ISBN 978-1-108-08281-5 Paperback

THE TOWER OF PELÉE

Photo. Heilprin.

FRONTISPIECE

The Tower of Pelée, seen from near the crater's edge, and from an altitude of approximately 4000 feet. Photograph taken on June 13, 1903, looking south by west, and from a distance of 700–800 feet. The height of the nearly vertical tower as it here appears is about 840 feet (the thickness of the base upwards of 500 feet), the sheer wall of rock rising out of a supporting cone or "dome," the summit of which considerably overtops the actual crest of the volcano. The left-hand face of the tower—that which is in shadow—is the face, turned to the east-northeast, which appears in Plate V., and shows the smooth side rubbed out by attrition. The rough northern face is scraggy, almost bouldery, through irregular breakages, and shows the effects of explosion-disruptions. A recent decapitation of the summit is distinctly indicated in the even transverse line which appears a short distance below the top. The vapor masses shrouding the base of the tower are almost wholly of steam—steam that is being forced by the volcano through the mass of the basal cone itself, and partly through the contact zone of the tower and cone. The aspect of the tower from this point, with the steam and ash-puffs and blue sulphur fumes playing about its base, was one of extraordinary magnificence.

THE

TOWER OF PELÉE

NEW STUDIES OF THE GREAT VOLCANO OF MARTINIQUE

BY

ANGELO HEILPRIN, F.R.G.S.,

OF THE
SHEFFIELD SCIENTIFIC SCHOOL OF YALE UNIVERSITY

LATE PRESIDENT OF THE GEOGRAPHICAL SOCIETY OF PHILADELPHIA,
MEMBER OF THE AMERICAN PHILOSOPHICAL SOCIETY, ETC.
AUTHOR OF
"MONT PELÉE AND THE TRAGEDY OF MARTINIQUE," ETC.

ILLUSTRATED

PHILADELPHIA AND LONDON
J. B. LIPPINCOTT COMPANY
1904

Published December, 1904

PRINTED BY J. B. LIPPINCOTT COMPANY, PHILADELPHIA

PREFACE

THE following pages, dealing with one of the most remarkable structures that have ever been described from the earth's surface, relate to what might properly be termed the middle period in the modern history of Pelée. The cataclysms of May 8 and August 30, 1902, had, it was thought, measured the full activity of Martinique's re-born volcano, and closed the particular chapter in vulcanology which it had opened. The construction of the volcano's extraordinary and unique excrescence has, however, once more shown how feeble may be the knowledge of phenomena that are ordinarily assumed to be fairly well measured, and placed before the investigator new problems and a new field for investigation which could hardly have been anticipated. The particular object affecting these problems for the moment no longer exists, but for that reason the investigation is not less interesting and important.

In his third visit to the island of Martinique the author was again made the recipient of the hospitality which the Clerc mansion at Vivé afforded. To the kind people, and now "old friends" of the estate, who made his investigations in the summer of 1903 pleasant as well as profitable, he is under lasting obligations, and he can but ill repay their generosity and goodwill in the expression of thanks which these few words convey.

ANGELO HEILPRIN.

NEW HAVEN, December 1, 1904.

CONTENTS

7

LIST OF ILLUSTRATIONS

THE TOWER OF PELÉE

I

MARTINIQUE REVISITED AND A FOURTH ASCENT OF PELÉE

Not quite a month after the first anniversary of the destruction of Saint-Pierre, I again set foot on Martinique soil. The silent city remained much as it was at the time of my last visit, nine months before. A little more ash had accumulated here and there, and some of it had been taken off elsewhere; but the ruins were the same battered, crumbling walls, unchanged save that they had gained in color through the washing off of the ash-mud that plastered and cloaked their vertical sides. In a few places excavations were being made to recover "treasure" or to locate sites, but the prowlers among the dead were few and what was recovered was in most cases insignificant. I turned over some rubble-masses beneath which "caked" and burnt papers were projecting, and found that I was dealing with a lesson in geology, and, strangely enough, with one that taught of volcanoes and volcanic phenomena—several pages of manuscript, possibly escaped from the Lycée or the Communal College, covered with teachings of Vesuvius, Cotopaxi and Etna (and of Pelée?). It may be that those papers were dictated by the impending storm of Pelée, but who can now tell? The fragment of one of the few books recovered from Saint-Pierre—whose precious brown pages I owe to a friend—deals likewise with volcanic phenomena. It is the *"L'Enfant du Vésuve,"* supplemented with a very full account of the destruction of Pompeii and with a carefully rendered translation of both of Pliny's letters.

One significant change had come over Saint-Pierre. It was no longer an absolute desert, for little colonies of ants and other insects were inhabiting the ruins and the land-snail had come to live with them. Green creepers and many plants with bright flowers here and there hung about the battered masonry, and from some of the old gardens rose up stocks of the *chou Caraibien* and the banana. And

11

even the few trees that had been left standing on the surrounding
heights, and thought to be dead, had sprouted out new leaves and
given a new sunshine to the landscape. Well up on the volcanic slope,
beyond the Roxelane, and quite to the Rivière des Pères, these signs
of returning vegetation were apparent, and on one side of the Roxelane itself everything was green. But, after all, it was more the immediate foreground that gave these signs of resuscitation, for farther
beyond, and below the hanging volcanic cloud, the grays were as gray
as ever, and the valley of the Rivière Blanche, choked with the immense
amount of débris that had been thrown into it, was white like snow
with the new ash that is periodically being swept over its course.

At Morne Rouge, which fell in the storm of August 30, not a house
remained inhabited. The beautiful church under whose partially lifted
roof good Père Mary had sought refuge for nearly his last hours, still
stands with its foot in the ash. My attendant climbed into the belfry
and tolled the bells that hung uninjured from the posts. It was the
voice in the wilderness, for there were none to listen to it but ourselves.
Perhaps far away on the hill-sides, where specks of cottages appeared
in the surrounding green, some may have recognized the beautiful
resonant tones.

The exquisite woodland that previous to August 30 bordered most
of the road between here and Ajoupa-Bouillon, stood out now as ragged
tree-trunks, spectres in the destroyed landscape, with naked arms and
upturned roots, begging, as it were, from the new sunlight that surrounded them. Here and there the eye fell upon the returning fronds
of the tree-fern and clumps of bamboo, on the melastome and broad-
leaved heliconia; but they were merely visions of what had been
before. Miles away over the landscape the eye still caught the images
of the wreck and ruin which that fearful blast of the late August day
had wrought. Mountain slope and valley were swept alike, and even
upon the ascending heights beyond the Capot the scars of destruction
remained luminously implanted. A wayfarer at Ajoupa-Bouillon, who
had lost all that was dear to him, pointed out to me a spot on the
open road where five of the village inhabitants, who had taken refuge
under a culvert, succeeded in weathering the storm, while almost everything about them was hurled to annihilation. I myself noted with
considerable interest that many of the wayside shrines, whose faces
were turned somewhat off from the direct path of the tornadic storm,
retained their contents almost undisturbed. The goblets, though filled

12

with ash, were intact, and the images were largely so. At this point, evidently, the destroying blast had lost much of its force, but on other lines, for far distances beyond, there would seem to have been little diminution to its power.

From another of the village inhabitants I obtained a graphic account of the awful hours that preceded the fatal explosion,—hours that followed immediately upon the time when my own little party left the upper volcanic slope,—of the black, but luminous night, the deafening roar of the volcano, and the final and terribly swift oncoming of the destroying cloud. I was especially interested in his description of the electric characteristics of this cloud, the short and rapid discharges and incessant crackling being, as the narrator stated, only comparable to *feu d'artifice*, an observation that had already before been made in connection with the Pelée cloud of May 8, and which only further confirmed me in the belief I have elsewhere expressed that electric discharges must have played an important part in the destruction of human life, both here and at Saint-Pierre.

As on my former visits, I made my head-quarters on the northeastern side of the island, to windward of the volcano. The great sugar-plants of Vivé, Leyritz, and Basse-Pointe had once more set their wheels going, and it seemed that for some time at least a cheerful life might again replace the dismal depression which months of despair had brought on. The old score against the volcano was for the moment wiped out. The proprietors and *gérants* had tired of the uncertainties of volcanic action, and between abandoning their estates absolutely or transporting what little could be transported elsewhere, and remaining to face possible death from an uncertain eruption, they chose the latter course, as perhaps most persons in their unhappy position would also have done.

My window in the capacious Clerc mansion at Vivé opened out upon a clear prospect over the summit of Pelée, and at times when there was little "volcanic cloud" hovering about, which was much less often the case than the reverse, it gave a fine view of the surmounting giant obelisk. Several times during the nights of my stay I was tempted to pass to the window and follow with a powerful glass the activities of the volcano. There was, however, little beyond landscape prospect to reward the search, except on the evening of June 12, when the base of the tower, in its southwest corner, was brilliantly luminous, being fed with volcanic fire through the interstices and rifts that pene-

13

trated the column. It was a beautiful spectacle. The fiery form appeared shortly after sunset, and it prompted me to make an ascent of the mountain on the following morning.

On June 13, in company with one of the officers of the French Scientific Commission, I made my fourth ascent of Pelée. The passing night promised everything. A few high clouds hovered about the blue and receding mornes that stretched off towards Carbet, but over the volcano itself there was nothing, and the great obelisk, its base fiery red with molten lava that was being poured into it, stood out in bold relief against the green-blue western sky. We left our quarters early, so as to gain upon the clouds that viciously gather about the summit; but the clouds had preceded us, and already at the breakfast hour, by which time we had reached the former summit, everything was wrapt in cloud and mist, and little was visible beyond ourselves. We succeeded in steering a course across what had before been the basin of the Lac des Palmistes, and in a few minutes stood upon the edge of the great crater. Everything was gray within,—not silent, however, for avalanches of rock were being precipitated and tumbled about in ruthless manner, and an occasional ominous roar told that the spirit of the mountain had not entirely departed. For the better part of six hours we vainly strove to penetrate the sea of cloud and fog that hung ahead of us. Each coming gust seemed to give us the chance for which we were waiting, but the rising crater-vapors kept the basin full, and even under a clear sky they allowed only "memories of a landscape" to escape. Although in no way unbearably hot, I found the crater rim uncomfortably warm and humid; it seemed to me more so than on my earlier visits. The actual temperature was only 85 degrees, however.

We found the entire depression of the Lac des Palmistes filled up and over by volcanic ejecta,—sand, pumice, and boulders, perhaps in greater part the product of the August 30 eruption. There was now a gentle and nearly uniform slope up to the crater-border, and what remained of the Morne de la Croix was hardly more than a rising knoll or knob. The split-boulders, or what have been called "bread-crust bombs," were very numerous, measuring in all sizes from small balls to masses two and three feet in diameter, and were lying freely scattered around. What surprised us greatly were the swarms of a bottle-green coccinella that had made a home on the summit. The tiny insects appeared to be about in myriads, and in an

14

instant almost our clothing was covered by hundreds. What they found on the barren summit to attract in this manner is one of those mysteries of nature which we found impossible to fathom. The species, which we failed to determine, was probably the common form of the island. More singular yet, we came across a stray bull-frog of large size, whose excursion to the top summit was equally inexplicable.

From the crater's edge we could at times look down to the very bottom, but the shifting vapors were such as to give us only flashing vistas, and for many hours we could frame no distinctive picture of what we saw. Steam jets were issuing at many points, and with these curled out the blue puffs of sulphur. In a very rough way I estimated that the depth below where we were standing could not have been less than 300–350 feet, which is very nearly twice what had been assumed by some of the observers of the French Scientific Commission. A later photographic measurement would seem to confirm my determinations.

The clouds continued to move, to break, and to unite, and for a long time it seemed as though we should be obliged to miss the object of our search entirely. There were brief spaces of atmospheric lucidity, but they were in the wrong quarter for us, and only showed up with transcendent beauty the landscape that was back of us and down the mountain. We were quite close to the edge of the crater, hardly three feet intervening, and vainly peered through the sea of mist and vapor to obtain a single glimpse of the avalanches of rock that were being tumbled down ahead of us, seemingly in space and from space, whose roar went out like the distant flow of thunder. We listened and heard everything; we strained our eyes and saw nothing. *Quelle mauvaise chance!* uttered my associate, and I echoed it most heartily.

Shortly before two o'clock the opportunity for which we had so impatiently waited seemed finally to arrive. Clouds and vapors died down to one side, and the great tower, its crown hanging at a dizzy height above, began to unfold. Piece by piece was added to it—purple, brown, and gray—until at last it stood abreast of us virtually uncovered from base to summit. "Look!" I shouted to my companion, and my words failed me for the magnificence of the view that presented itself. The spectacle was one of overwhelming grandeur, and we stood for some moments awed and silent in the shadow of this most impressive of mountain forms. Nature's monument dedicated to the 30,000 dead who lay in the silent city below, it rose up a huge monolith, 830 feet above the newly constructed summit of the volcano, and

15

5020 feet above the Caribbean surface,—a unique and incomparable type in our planet's wonderland.

We spent about two hours and a half on the summit after the first rifting of the clouds, and had thus a full opportunity to study from most sides, even if not absolutely close at hand, the general characteristics of the giant tower and of its setting. M. Guinoiseau, who had at this time made the ascent of the volcano perhaps more than twelve times, was as enthusiastic over the scene as I was myself; but he reported that the volcano was in an unusual state of eruptivity, a not exactly comforting assurance to the plain folk who had already come to know the burning mountain. However, we saw little to disturb us in our studies, and it was rapidly nearing five o'clock when we began our descent of the cindery slopes. Shortly before seven o'clock we again entered the hospitable portals of the Usine Vivé.

II

No other name, it seems to me, more appropriately conveys the picture of the giant core of rock, nearly 1000 feet in height at the time of its greatest development and 350–500 feet thick at the base, which Pelée had bodily lifted and pushed out from its summit during a period of a full year and more. This extraordinary obelisk of lava, like a veritable "Tower of Babel," whose apex at the time of my visit, the middle of June, 1903, reached a position 5020 feet above the sea, transfixed the newly-formed cone of the basin of the Étang Sec, and rose to all purposes vertically above it, the two structures, products of the eruptions beginning in April, 1902, having a full height of approximately 2300–2400 feet. As seen from the east-northeast, or the quarter of Assier and Vivé, it presented the aspect of an acute pyramid; seen from the south or southwest it gave the appearance of a conical spire, complicated by secondary spires, needles, or fingers, and showing a split or indented apical summit; while from the northeast and north it rose up a gigantic and nearly parallel-sided tower or fortress. From whichever side seen, it was an object of sublime magnificence; and in its condition of vapor clouds blowing out from its base and from the cone that supported it, with blue sulphur smoke curling its way along with these, it presented a spectacle of almost overwhelming grandeur and one of terrorizing effect which could hardly be matched elsewhere. None of the grand scenes of nature which I had before seen—the Matterhorn, the Domes of the Yosemite, the colossus of Popocatepetl soaring above the shoulder of Ixtacci-huatl, or the Grand Cañon of the Colorado—impressed me to the extent that did the view of Pelée's tower, from the crater-rim, on the afternoon of June 13.

The tower was arched slightly in the direction of Saint-Pierre,— *i.e.*, towards the southwest, where the surface was scraggy, and apparently scoriaceous or slaggy, the result, doubtless, of the numerous basal eruptions which took place at or near the point of contact with the supporting cone. The surface on the opposite side—that turned towards Assier—was, on the other hand, smooth, almost polished in places, and longitudinally grooved from base nearly to summit. This smoothness and graving of the surface were certainly due to attrition against the encasing rock or "mould" which formed the wall to the channel of exit, and the curving over of the mass to one side would seem to point to extrusion from beneath a somewhat vaulted or curved casing. One could well compare the structure and its method of escape to a core of paint issuing by pressure from an oil-tube. The general surface-covering was in color ruddy gray, brown and purplish in part, but on the smooth face it was nearly white, a condition probably in some way associated with the rubbing on that side.

As to the fundamental and inner construction of this remarkable volcanic appendage our knowledge remains in a measure conjectural. As seen with a powerful glass from a point of nearest approach, perhaps 700 feet, the rock appeared "burnt-out," like a furnace-product; and the noise given out by the falling particles and boulders was generally like that of falling clinkers, which might have led to the supposition that the mass was on the whole cavernous. But its rigid adherence and resistance to a prodigious crushing strain lend little countenance to this view. The noise from the more imposing discharges of dejecta was like that of rolling thunder, at times barely distinguishable from the roar of the volcano itself, and could hardly have been produced otherwise than by the avalanching of compact rock.

It has been surmised, or at least suggested, that the interior of the tower might have been hollow, with fluidal lava, hidden from view by the massive outer walls, contained within. This condition is not conceivable. Had such a chimney with an enclosed flowing magma really existed, there would certainly have been lava overflows at one time or another. On the other hand, that the tower was rifted and had irregular passages through it or through parts of it, into which lava was at times injected, is certain; and the members of the Lacroix mission on more than one occasion noticed areas and lines of incandescence in the basal portion of the core, which they associated with flowing lava-masses. On the night preceding my fourth ascent of the volcano, June 12, 1903, the southwest base of the tower was resplendently luminous, made so either by actually rising lava or by a partial remelting of that portion of the structure. From a distance of a few miles, whence this magnificent spectacle was seen, my powerful glass failed to determine which of the two conditions existed,—a matter of little consequence, as in either event molten lava was in close association.

That some of the rifts completely traversed the tower from base to summit, I had the opportunity fully to satisfy myself, for on the morning of June 15, when skirting the northern and western shores of the island, a thin steam-pennant could be seen to be continuously issuing from the apical summit; in other words, the volcano was gently "smoking" at the top. The issuing vapor was perfectly white, and it seemed to carry little or no ash with it. From the same apical summit a number of incandescent balls are reported to have been shot out on the night of March 26, 1903.

The ascent of this remarkable core of rock, the general nature of which was first determined by Prof. Lacroix, was due to processes similar to those which produce the outwelling of lava in the ordinary form of volcanoes,—i.e., to interior volcanic stress. Despite its colossal dimensions, the tower was heaved bodily upward, receiving new accretions of matter almost entirely from below. The most cursory examination of the relations existing would immediately point to this form of growth and development, but the carefully conducted angle-measurements and observations of contour made by the representatives at two stations of the French Scientific Commission leave no possibility for doubt in this matter, and they further furnish us with data touching the rate of growth. The consideration of the depth to which this giant monument descended solid into the volcano would be interesting were there any way of reaching the problem, but for the present there would seem to be none such. It is perhaps enough to say at this time that this depth must have been considerable, otherwise the column could not have stood through the exploding condition of the mountain; the depth, again, may have been very great. On the other hand, the problem cannot lose sight of the fact that molten or incandescent lava did at times rise quite to the level of the insertion of the monument in its base.

It is a matter of some importance geographically to know when this great tower of rock first appeared and to ascertain through this fact its relation to the great eruptions of May, June, July, and August (1902). Prof. Lacroix, in an article published in the *Dépêche Coloniale* (April 30, 1903), states that the basal cone of the volcano had been terminated by a needle since the middle of October, and presumably this is about the period when it was first seen by him. But there can hardly be a question that its formation or first appearance was of much earlier date, for on August 24, 1902, almost a week before the second death-dealing eruption, a vertical (although comparatively short) needle was distinctly seen by me from the southwest side, and it appears in my photographs taken on that day. Indeed, I remark in my report,* that it seemed to me likely that the two glowing masses of fire which shone down from the summit, like red beacon-lights, in the morning of August 22, emanated from the two (incandescent) horns that capped the summit of the mountain. One of these protruding masses, or "horns," as I have called them, was seemingly set at a broad angle to the other.

In an earlier report on my observations and experiences,† published shortly after my return from my first visit to Martinique, use is made of a drawing of the crater by Mr. George Varian, an artist associate who was with me when we first reached the rim of the still very active crater, and whose extreme faithfulness in the delineation of nature I frequently had occasion to admire. In this drawing a great core of rock is made to appear centrally in the crater and rising somewhat above the crater's rim. In my own description (p. 365) I refer to these points in the crateral structure as "the central core of burnt-out cinder masses, topped by enormous white rocks, whose brilliant incandescence flashed out the beacon-lights which were observed from the sea some days after the fatal 8th, and even at our later day illumined the night crown of the volcano with a glow of fire." When at that early day we stood on the crater's edge, the activity of the volcano was still such that we could obtain but momentary glimpses of the interior of the crater and of the crater-walls, and it was impossible to shape constructively the relations of the parts as they passed before us in fleeting shadows. After seeing one of my own photographs and the photographs of investigators who were on the volcano after I had left it, I became doubtful as to the accuracy of Mr. Varian's drawing, the more so as it depicted a structure that could not be brought into relation with any known volcanic feature, and in my later publication I thought fit to omit the illustration. There is no doubt in my mind at this time that the sketch of my artist associate was an accurate one, and that "the central mass of jagged white

* "Mont Pelée and the Tragedy of Martinique," p. 163.
† McClure's Magazine, August, 1902.

20

rocks'' was already as early as May 31 the embryo of the great Pelée tower. That it did not survive into the later day is certain, for on June 20, when Dr. E. O. Hovey took his photograph of the central cone, it no longer appeared.* It was probably overthrown in the forceful eruption of June 6. On the other hand, it is certain that it reappeared within the period of a few weeks, for it is distinctly shown in a photograph taken on July 6, which is published by Dr. Jaggar.† It should be said that on May 31, 1902, the sound of falling "clinkers" (rocks, etc.) was precisely that which we heard on June 13, 1903, emanating from the falling and exploded débris of the obelisk.

The giant tower at the time of my visit reached an absolute elevation above sea-level of 5020 feet, the determination made by M. Guinoiseau from Assier, with which a less accurate Abney-level measurement made by me from Morne Rouge closely agrees. Its height was on May 31 5200 feet, but it lost on that day through breakage 180 feet of its summit. It frequently underwent partial decapitation, and the form was thus largely disturbed, the summit or apex particularly suffering. During the four days preceding June 15—within the period of my latest visit to the volcano—the ascent, as determined by angle measurements made by M. Guinoiseau at Assier, was six metres; in the eight days preceding June 7, ten metres.‡ However incredible such a rapid rise may appear, the facts that are presented in the first period of the tower's history are yet far more imposing than those of this later day, and of a kind to impress upon the observer, in a wholly exceptional way, the sense of sovereign grandeur of nature. Were it not for the immediate object placed in full view, there would be few, even among extreme cataclysmists, who would be prepared to believe that for a period of a month or more so gigantic a structure as the Pelée tower could have been heaved up at an average daily rate of from 20 to 25 or even 25 to 30 feet. As has already been stated, the first appearance of the forming tower or spine as noted by the French Scientific Commission was on or about October 15, 1902, and by the close of November, despite partial breakages of the summit, this extraordinary structure had risen to approximately 1500 metres (4920

---

* Bulletin Amer. Mus. Nat. Hist., xvi., pl. 44, Fig. 2. See also a more recent paper by Dr. Hovey in the Amer. Journ. Science, Oct., 1903, p. 271.

† Amer. Journ. Science, Jan., 1904.

‡ M. Giraud, in L'Opinion, of Martinique, June, 1903.

feet) above sea-level.* Of this total height about 800 feet fell to the tower alone, a rise, therefore, of this amount, with breakages, in from 35 to 40 days.†

The history of the tower since the early days of December, 1902, is one of frequent breakages and of an almost continuous repair, so far as mere elevation is concerned, resulting from progressive and virtually continuous upheaval. Marked changes of outline, particularly as seen from the east and the southeast, followed the major disruptions, and to that extent that it has been made difficult to harmonize even photographic views taken at different times from nearly identical positions. In the first week of December the height of the tower was lessened by 60–70 metres, but this loss was made good in a very few days, and despite successive losses the tower on December 16 rose to within 70 metres of its greatest height. Through, or accompanying, the rather severe eruption of January 25, 1903, there was a further loss of 30 metres (at first reported to be 250 metres), and at this time it was observed that the volcano was capped by two needles.‡ This interesting fact is significant in its relation to my own observation that two "horns" or needles projected from the newly-formed cone on August 24 of the year previous. On March 13, the date of Lacroix's departure from the island, the tower had risen to 1568 metres (5143 feet), overtopping the remains of the Morne de la Croix by 1009 feet; § but at the end of two weeks, in the eruption of March 26, when it is reported that incandescent "balls" were shot out from the actual apex of the tower, it again lost 25 metres (82 feet).¶ Seemingly the extreme height that was reached by this extraordinary volcanic structure was almost exactly 5200 feet on May 30–31, 1903. At that time, as I was informed by M. Guinoiseau, another eruption re-

---

* Comptes-Rendus, Dec. 1 and Dec. 29, 1902.

† Major W. M. Hodder, of the Royal Engineer Corps, from observations made at Morne Fortuné, on the island of St. Lucia, nearly 60 miles in a direct line from the cone of Pelée, determined the absolute height of Pelée on Nov. 26, to be 5032 feet, about 100 feet greater than was reported by the French Commission (communication to Dr. E. O. Hovey, Amer. Journ. Science, Oct., 1903). The low angle of measurement probably makes this determination less accurate than the French.

‡ Comptes-Rendus, Feb. 16, 1903.

§ Dépêche Coloniale, April 30, 1903.

¶ The elevation of 5143 feet is almost exactly that which was found by Hovey on March 25: height above crater-rim, point of observation, 1174 feet; height of crater-rim. 3969 feet above sea-level.

moved 180 feet. There can be no doubt that had there been no apical disruptions the tower would have reached a full thousand feet higher.

On comparing my photographs taken from the crater-rim on June 13, 1903, with those of the French Commission and others, especially the very beautiful ones of Dr. Hovey, one is struck with the remarkable changes of outline which the tower had undergone,—changes that could have resulted from breakage alone, except perhaps at the immediate base. From no point of view on the old basin of the Lac des Palmistes could I obtain a picture of the tower that was more than suggestive of what appears in the photographs of Hovey taken eleven weeks before (on March 26), and which illustrate his article on "The New Cone of Mont Pelée." * Equally "irreconcilable" are still earlier pictures which I found in possession of local photographers in Fort-de-France.

The numerous breakages and decapitations which the tower undergoes naturally suggest that the materials of its construction, while sufficiently solid and resisting to permit the mass to hold its weight, cannot well have had the consistency of granitoid or plutonic rock or of lithoid lava,—at least not in its outer parts. I suspect, and it has already been stated before me, that much of the exterior at least was pumiceo-vitreous in texture, and sufficiently so as to permit of easy dislodgments, even as the result of jarring alone. It is certainly a curious fact that almost every moderately severe eruption threw down a portion of the summit, besides at different times opening great longitudinal fissures. Such a cleft was opened by the eruption of November 18, 1902, and through it a slice of the tower measuring nearly 300 feet in height (90 metres) was removed. Other fissures followed rapidly in the early part of 1903, producing those modifications in the tower which Major Hodder has likened to the change from a "huge lighthouse" to the form of a church-steeple. The surface aspect of a large part of the northern and western faces of the tower gave clearly the picture of a slaggy (cavernous or pumiceous, it might be called in a certain sense) structure, or at least of something that was not compactly solid; and the lines of fracturing would seem to reveal something of like nature. At the same time, too much dependence cannot properly be placed upon this surface-appearance, as the limits of size for individual parts were apt to be misjudged in the vastness of the whole,

---

* Amer. Journ. Science, Oct., 1903, cuts facing p. 276.

and to give impressions from which false conclusions could easily be drawn. It was more than regrettable to me that at the time of my visit in June, 1903, the condition of activity was again such as to prevent me from descending into the crateral hollow and of examining the numerous blocks that were almost continuously being dislodged from the tower.

In considering the question of the disruptions and summit-falls or decapitations of the tower, the fact must not be lost sight of that the entire height was at times penetrated by steam, which rose not through a central or permanent chimney, but along one or more rift-passages. As I have elsewhere noted, the ascending steam was observed by me, during two hours or more, to pass out distinctly from the actual apical summit in a delicate line of pennant.* It may be that it was precisely this tower-contained steam which, with additional force given to it at times of special eruptions, was responsible for the lofty disruptions, as well as for the dislocations on the upper flanks.

While we are thus not in a position to state exactly what was the inner construction of the tower, it is perhaps not unreasonable to assume that it was rigidly solid (whether of a lithoid or glassy, or obsidian-like, structure) in its *major* part,—as the polished side directed towards Assier would indicate, and the giant rock-masses hurled into the valley of the Rivière Blanche almost certainly prove,— and that a cavernous, slaggy, or pumiceous exterior surrounded the more solid interior core. But whatever this structure may have been, it did not affect the imposing character of the object or of the phenomenon which it portrayed.

It will naturally appear to all who have reflected upon this new manifestation of volcanic activity that the power to lift or even sustain so gigantic a structure as this tower, with a cubical content (even if less in weight) equal to that of the Great Pyramid of Egypt, must have been prodigious. But the problem from the purely geological side is merely that of the normal volcano pushing up its great cone of molten lava, with this difference, that in the Pelée uplift the element of friction enters as an important factor in the calculation of the dynamics of the lifting force. What may have been the value of the differential, unfortunately, in the absence of knowledge regarding the

---

* Dr. Hovey appears to have been less fortunate in his observation, for he remarks that no steam was ever blown out from the top.

fixation of the tower, cannot be stated, nor even approximately hypothecated. Our present knowledge of volcanic phenomena, indeed, does not even permit us to make a comparison between the lifting force of the low volcano and that of the lofty cone, two or more miles in height.

The cone that supported the tower, or rather through which the tower passed, and which remains to-day, has been built up entirely since April 23, 1902. At the time of my latest visit (June, 1903) it overtopped the general summit of Pelée by about 200 feet, and had therefore an absolute height of some 1600–1800 feet. It is to this structure, implanted upon the basin of the Étang Sec, that I refer in my earlier reports as the "new fragmental cone." Dr. Hovey refers to it in the same relation. The exterior seems to me to have always been in great part a mass of débris, volcanic ejecta of all sizes, through which steam was puffed at numerous points. Solid lava ridges protrude (or protruded) through it, and give to it, especially in the southeast, a ribbed structure. The base occupies almost the entire floor of the former tarn-basin. Prof. Lacroix, who enjoyed unusual advantages for the study of this seemingly normal volcanic structure, asserts that the same is not a true fragmental cone, but a dome or monticule of lava without crateral opening, formed in the manner of the famous pre-crateral dome of Giorgios, in Santorin, of 1866. It will be recalled that in the formation of that interesting structure there was an upwelling of highly viscous lava, which simply accumulated in an irregular bouldery mass about the opening of the volcanic chimney. At a somewhat later, although still early, day a crater opened in this monticule or dome, and from it flowed out streams of molten lava. I am not convinced that the early stage of the Pelée cone, however it may have become modified later on, was of this structure. The photographs that were taken prior to August 30 show, nearly all of them, where the summit is at all visible, a truncated top, a form absolutely like that of the normal crater-cone and as much unlike the monticule of the Giorgios (or Puy) type. This is beautifully shown in the photograph taken by Dr. Hovey on June 20; and equally so in my own photographs taken from the west side on August 24 (1902), which depict the volcano "smoking" directly from this summit chimney. It seems to me more likely that the conditions of Santorin have been simply reversed in the case of Pelée: a fragmental cone was opened first, and only later became plugged by the rise in it of what ultimately became the tower. Indeed, it would appear that the plugging went on as a process continuously with the making of the cone, troubling the volcano in its workings, but yet not so far obliterating the structure that held it as to obscure its relations. The fact that the tower passed bodily through the cone is in itself evidence of a kind supporting the view of a crateral cone. That the cone or dome at a later period acquired more or less of the structure that Lacroix ascribes to it, there can be no question.

In their local setting the obelisk and its supporting cone occupied the virtual centre of the basin of the Étang Sec (the erupting crater of 1902), the floor of which in June, 1903, had been brought up by infilling to within 300–350 feet of the summit of the surrounding wall of the caldera. The width of the space separating the top of the cone from this wall was roughly estimated to be 200–250 yards, excepting in the west, where the cone had coalesced with the slope of the Petit-Bonhomme (Ti-Bolhomme). Basally the cone almost united (did unite in places) with the slope of the caldera wall, the discharging débris constantly narrowing the space between the two structures. During the time of my observation the cone was in a fairly active condition, blowing out steam at numerous points; and there could be no question that these issuing steam-puffs came directly from the interior, and were not secondary explosions emanating from the covering débris. According to M. Guinoiseau, one of the observers of the French Scientific Commission, the activity at this time was particularly accentuated, greater than it had been at any time since the month of January preceding.

In my "Mont Pelée and the Tragedy of Martinique" I have given a fairly extended survey of the Pelée crater, to which there is little to be added beyond what is contained in the preceding pages. It might be noted that it was (and is) of the caldera type,—i.e., with its partially encircling walls trenchantly steep, almost or quite vertical in places,—and clearly showing the lines of stratification of the super-imposed fragmental materials (pumice, ash, etc.). These are traversed by the base or "dike" of the Morne de la Croix, whose andesitic mass could be followed by the eye virtually through the entire height of the wall (300 feet high).* Into this caldera, the basin of the former

---

* It should be stated that some of the observers of the French Scientific Commission were disposed to give a considerably less height for this wall (the depth of the crater) than I have here given.

Étang Sec, was implanted the new cone, with its great transfixing obelisk. It has been remarked that on the western side the cone had united with the basal processes of the Ti-Bolhomme; elsewhere it was surrounded by a V-shaped valley (*rainure*), the top-width of which was roundly 600–800 feet.

There is nothing in the configuration of the mountain to give countenance to the extended caldera-form, with the mass of Pelée as a centrally rising polygenetic cone, which Stübel has fancifully constructed from the large French map of the island of Martinique, the contours of which have only distantly approximate relations to the actual relief of the land. There is on the north side of the crater, but absolutely on the summit of the mountain, the remains of an ancient crateral wall, which Dr. Hovey has already likened to the Somma wall of the Atrio del Cavallo, but it bears no relation to the contours shown on the French map. It is a short way back from the edge of the present crater, and its rocks are partially columnar.

III

THE AFTER HISTORY AND NATURE OF THE TOWER

IN the destruction of Pelée's giant tower, and the entry of the volcano into a new condition of activity, one of the most remarkable features of the earth's surface has disappeared; and while there are indications that the structure might be replaced by something similar, its removal takes from the eye of the geologist an object illustrating a unique phase in the history of volcanic phenomena. It is wholly likely that at some earlier period of the earth's history structures have been developed similar to the Pelée tower, but it has not been given to the geological observer to study their formation or even to identify their relations. Hence the significance of the opportunity for new studies which have latterly been presented.

The systematic destruction of the great core of rock began in the early days of July, 1903, and was accomplished with despatch, so that by the middle of that month there was a loss to the summit of nearly 400 feet, and before the close of the second week in August of a further 100 feet. During this period of destruction, and for weeks afterwards, the activity of the volcano was very pronounced, and discharges of the *nuées denses* (the name given by the French Scientific Commission to the descending black clouds which were thought to be similar to the cloud that destroyed Saint-Pierre) were frequent, not alone along the valley of the Rivière Blanche, but also directly towards Prêcheur (August 20–21, 27–28, September 4–6, etc.) and across the former basin of the Lac des Palmistes (August 20–21, 28–30, September 11–12, 13–14, etc.) The greater number of the discharges continued, and seemingly still continue (November 2–3, 3–4, etc., and at different times in 1904), along the valley of the Rivière Blanche, taking the course of the hurricane-blast of May 8, 1902.

Coincidently with this destruction of the tower and the return of Pelée to a condition of fairly forceful activity, it was observed by the French Scientific Commission that the conical base upon which the tower was implanted—or, rather, through which it passed—was itself undergoing marked modification, being forced up in the manner of a dome. Effusions of viscous lava were adding to its mass, breaking through the confines here and there, and solidifying before there was an opportunity for a free flow. At other times the growth of the "dome" was seemingly merely an outward-swelling or expansion (intumescence), an ebullition, resulting from steam-pressure and the accretion of lava rising from below. Whether thus formed by exogenous additions or as the result of endogenous accretions, the growth of the dome, despite the not insignificant and sometimes very pronounced losses to its mass following upon almost every larger explosion, was remarkably rapid. During the ten days preceding August 17, as we are informed by Prof. Giraud, the dome had gained in height 88 feet (27 metres).* During somewhat less than eight days, immediately preceding August 27–28, the gain in height was 164 feet (50 metres); and between August 26 and August 30, 98 feet (30 metres); † showing an average daily increment of 24½ feet, closely correspondent with the rate of growth of the tower in its earlier period. The phenomena attending the growth of the dome were those of the general eruptions: the evolution of the great volcanic steam-ash cloud, rising at times to 2000, 3000, and 4000 metres above the summit of the volcano; loud detonations, frequent discharges of dust and boulders, and the more forcible explosions of the "black cloud." During the eruption of September 9, which lowered the dome 15 metres, the *nuage dense*, following the course of the Rivière Blanche, reached the sea in five minutes, thus repeating the history of the early period of Pelée's activity. In an earlier eruption, September 3, when the dome lost 30 metres, a similar cloud reached the sea in seven minutes. It is significant that at about the middle of September these clouds, instead of following the usual downward course, now in the main ascended vertically.‡ On September 15–16 such a cloud rose to the extraordinary height of 7000 metres.

During much of the period here noted parts of the dome appeared brilliantly incandescent, some of the luminous points being fixed for a number of consecutive days. On a few nights when observations were permitted, nearly the entire surface of the dome appeared as if in a

* E. O. Hovey, Science, Nov. 13, 1903, p. 633.

† Giraud, L'Opinion, Martinique.

‡ Giraud, La Colonie.

glow of fire, and brilliant reflections were thrown upon the clouds overhead. The trains of boulders discharged from the dome with almost
every forcible eruption were also frequently incandescently luminous.
It would seem that by the first of October the new dome had actually
been constructed to a height slightly exceeding 500 feet, doubtless
enclosing within itself a considerable portion of the lower moiety of
the destroyed tower, if, indeed, it did not again bring it to a condition
of molten fluidity. The most rapid development of this remarkable
structure appears to have been on August 30–31, when, as reported by
M. Giraud, the rise in a single day was 78 feet (24 metres).*

It is interesting to note in connection with the construction of
this remarkable crateral dome that it was (and is?) accompanied by
new extrusions of solid "turreted" matter, acicular processes or obe-
lisks appearing at different times in two or more parts of its summit.
Thus, in the early days of September (1903), the observers of the
French Scientific Commission noted that the dome was terminated by
an aiguille rising from its northwest part, which needle on September
7 rose nearly 10 feet. On September 9 this new growth acquired an
additional 6 metres, and between September 10 and 12 a further 8
metres. A second process was at a later day extruded through the
southeast portion of the dome, and its fortune, as well as that of the
earlier one, partook of the same vicissitudes of construction and destruction which marked the history of the original great tower. On
October 20 it lost 5 metres. There seems to be at this time no way of
ascertaining the precise relations existing between these newer structures and the basal portion of the first formed and partially buried
tower; nor can it be told if any structural relation in fact exists,
although I strongly suspect that it does. But that the newly appearing structures were in themselves of no mean significance is proved
by the observation that on November 25 one of the towers lost 30
metres of its height as the result of the eruption of that day.† At that
time the greater part of the dome was incandescent.

In comparing the Pelée dome (not the towers or processes) with
recalling or resembling structures elsewhere, the geologist naturally
turns to the two or three anomalous types of cone or summit that have
become known for their departure from the form of the normal volcano. These are the domed cone, already referred to, appearing on

the island of Giorgios, in Santorin, in 1866; some of the Puys, as the Puy Chopine, of the Auvergne region of France; and (perhaps most remotely) the pyramidated tops of volcanoes which Stübel has described from the equatorial Andes. The last-named structures, however, so far as I am able to comprehend Stübel's work, are seemingly only physiographic monuments associated with the original making of the volcanoes, and have nothing in common with a later crateral discharge. They belong to the type of Stübel's monogenetic and not polygenetic volcano.

Perhaps a still closer approximation to the Pelée dome is to be found in the dome-structure, first noticed in 1895, and more fully developed in the spring of 1898, which Matteucci has described in connection with the more recent outbreaks of Vesuvius, and which has been so persistently denied by Mercalli. Matteucci's studies are recorded in a number of very carefully prepared papers,* which leave little room to doubt the substantial accuracy of the observations which they present. We learn from these reports that the dome (or *cupola lavica*) gained in one month (February 14 to March 15, 1898) 15 metres in altitude, the floor of the crater swelling up (intumescing) at the same time to 50 metres. The total height of the dome is represented to be 163 metres. Matteucci sees in this upheaval the combined action of a deeply planted mechanical force and of a superficial intumescence, and he does not fail to recognize the conditions which are thought to be associated with the making of laccolites.† There would probably be no impropriety in designating the Vesuvian structure "laccolitic," even if it represents no true laccolite.

That the Giorgios dome, the type of the cumulo-volcano, is essentially representative of the structure seen in the Pelée dome, as Lacroix has urged, seems undeniable; indeed, the question of differences would seem to resolve itself, so far as a direct comparison is made possible, almost entirely to one of not very important details. The greater or

---

* Sur les Particularités de l'Éruption du Vésuve, Comptes-Rendus, 1899, vol. 129, pp. 65, 66; Sul Sollevamento endogeno di una Cupola lavica al Vesuvio, Rendicont. Accad. Scienze Fisiche e Matemat. Napoli, 1898, xxxvii., pp. 285 et seq.; Se al Sollevamento endogeno di una Cupola lavica al Vesuvio fossa aver contributo la sol idicazione del Magma, Bollet. Soc. Geol. Roma, 1902, xxi., pp. 413 et seq.

† " Si tratta di uno Sforzo mecanico profundo e di una Intumescenza superficiale," Rendicont. Nap., p. 299.

lesser activities of the two volcanoes may account fully for these differences.

Until the activity of Pelée will have so far lessened as to permit of a closer study of the dome its full nature cannot be determined, perhaps not even to the extent of allowing us to say in how far, if at all, it is related to the hollow, oven-like forms which Dana and others have described from the Hawaiian Islands under the name of "driblet" cones, and of which Israel Russell has more recently given us exaggerated types from among the Jordan Craters of Oregon. One of these "ovens" measures 20 feet in height and 40–50 feet in basal diameter.* That the intumescing Pelée dome is at times largely hollow seems sufficiently established by the markedly diminished height which follows or accompanies eruptions of only moderate intensity. In many cases of such eruption there would appear to be a general collapse.†

However closely we may approximate the structure of the Pelée and other domes, the problem of the tower remains thereby probably unaffected, and we again search among geological reliquiæ for parallels. Nothing appearing among recent volcanoes, one is almost tempted to make comparisons between the Pelée tower and those giant stocks of lava which have long been recognized by geologists as "volcanic necks" and "laccolitic cores," and which are presumed to owe their prominent forms in the landscape to differential erosion of the land-surface. That some or many of these cores are only such resisting blocks overlooking an eroded land-surface cannot be questioned, but it is not so certain that all are of this nature, and some may well be of the type of structure which Pelée has presented in its extraordinary tower. One cannot resist the conclusion, even without the direct support of facts, that there must have been other towers before the one of 1902, and some of these ought to be preserved somewhere; but where?

Sir Richard Strachey ‡ calls attention to "plugs" of trap, said

---

* "Geol. Southwestern Idaho and Southeastern Oregon," Bull. U. S. Geol. Survey, 1902, No. 217, p. 52.

† Professor Russell, in the report referred to, presents an exceedingly suggestive illustration of a (pressure) "dome in recent lava," also among the Jordan lavas of Oregon; but it is held that the lava of this and similar domes was antecedently horizontal, and was forced up as the result of later pressure. P. 54, pl. xv., Fig. A.

‡ Nature, October 15, 1903, p. 574.

not to be uncommon, rising out of the Dekkan plateau, and which he believes to be the analogues of the Pelée tower. A sketch of one of these, made as early as 1839, is in its form certainly very suggestive. Another structure might, perhaps, also be brought up for comparison in this connection. I refer to the giant Devil's Thumb,* on the northwest coast of Greenland, marking the entrance to Melville Bay. As I recall it from a distant view of two or three miles, after a lapse of twelve years, and as it appears in sketch on the border of the Admiralty Charts, it has almost exactly the outline of the Pelée tower, rising up in supreme and almost isolated majesty to a height of 2350 feet. Unfortunately, however, we are not yet in a position to state if this prominent feature in the landscape is volcanic, or even if one of the vast basaltic areas of Greenland absolutely surrounds its base. The relief and conditions of the land would seem to argue against any form of erosional construction.

NATURE OF THE TOWER.

There remains little doubt in my mind that the tower of Pelée was merely the ancient core of the volcano that had been forced from the position of rest in which solidification had left it. The generally accepted view regarding its construction is that which was advanced by Lacroix, and which holds that the giant block was an active acidic (andesitic) lava whose viscosity was such as to permit of solidification while still within the chimney of the volcano, and whose movement posterior to extrusion was, by reason of this solidificaton, necessarily made a vertical one. There could be no free flow. This explanation appeals in its simplicity, and it is one to which I confess myself having been at first committed. There are objections to it, however. The form of the tower, and the fact that it rose through a supporting dome or cone, a portion of which was constructed of actually fluid or semi-fluid lava. are hardly consonant with this mode of construction. Rather under these conditions would one look for a simple cumulus or dome, and for that alone. A general and united solidification within the chimney of the volcano over a surface having a diameter of 350–500 feet or more, and accomplished so rapidly as to prevent all overflow, is difficult to conceive of. The slow cooling of lavas is in itself a further serious objection, for it hardly permits us, even under the special

---

* Not the more southerly one bearing the same name.

conditions of volcanic stress here presented, to postulate the solidification to the core of so vast a rock-mass in the short period of its existence. Geologists have long taught the lesson of the many years in which rigidly cooled lava-streams have maintained their "fires" within; and yet here the extinguishment of these fires is claimed for the period of a few weeks, or even days. It cannot be doubted that the tower was virtually solid to the core, and equally little need one doubt that its temperature was not such as to maintain a fluidal or semi-fluidal interior. Had the tower not been solid, or had it contained much incandescent fluidal matter, the numerous breakages, whether on the flanks or across the summit, which marked the tower's history, would have revealed these conditions many times.

Again, the general aspect of the tower-rock was not such as to suggest recently cooled and solidified lava. I have elsewhere referred to its slaggy appearance and to its recalling "burnt-out" cinder masses, the whole looking much like a furnace-product and wholly unlike recently cooled lava. This was remarked of what might reasonably be assumed to have been the same structure as early as June 1, 1902, when the coming of a tower was not even suspected. With the solidification of an erupting fresh lava, while the outer coat would almost certainly be measurably scoriaceous, the great inner mass, unless parting with its gases in a manner that has not heretofore been observed or known in volcanoes,* could hardly be other than rigid rock, and one much more capable of resisting repeated destruction than was the rock of Pelée's tower. Nor, indeed, from this type of rock would we have obtained that clinkery sound which accompanied the numerous disruptions and falls of material from the tower.

Other objections to the commonly received view regarding the construction of Pelée's remarkable tower might be urged, but they would add little to a discussion whose premises are so difficult to reach as they are in this one.

In assuming the tower to have been an ancient neck-core which under enormous pressure had been lifted from its moorings, we at least require no condition that is not generally provided for by volcanoes. There can be no objection to postulating the existence of such a core here, as in other volcanoes; and if existing, there would seem

---

* See a paper by G. K. Gilbert on a (thought) possible construction of the Pelée tower, in Science, June, 1904.

to be no reason why, under the gigantic force of Pelée's activity, it should not have been dislodged and pushed bodily outward. The reaction upon this contained mass of accumulating heat, and the infusion into it of steam and flows of new lava, would help to explain the ''burnt-out'' and scraggy look which from the first had been a characteristic of the tower-rock.* This view of the nature and extrusion of the giant monolith would at the same time satisfactorily explain its isostatic condition and do away with the necessity of formulating new laws or conditions governing the rapid cooling of lavas. Indeed, in assuming the presence of this giant core in the throat of the volcano, blocking it and preventing the free escape of the impounded gases, it becomes much easier to understand the violence of the explosions which have marked so many of the Pelée eruptions and the disruptions that so repeatedly wrecked (more particularly) the southwestern base of the tower,—the side directed to Saint-Pierre or the valley of the Rivière Blanche.

To the objection that might be made that no similar extrusions have characterized the outbreaks of other volcanoes, it is not difficult to furnish the answer that they have not provided a tower of any kind either. The fact is that violent volcanic eruptions have been only sparingly studied, and few observers have been sufficiently fortunate to be on the field of activity at times when the earlier phenomena of an eruption could be profitably noted. There are, doubtless, many facts connected with the physics of the opening of a volcanic mountain which have heretofore escaped notice, and some of these may have been directly allied to the greater facts which Pelée itself has presented. The extrusion or lifting of giant solid masses by volcanoes is not, however, an absolutely unknown fact. Abich, as far back as 1882,† described the cliffs of limestone and marble which form an essential part of the centre of the crater of the ancient volcano of Palandokän, and which he unhesitatingly assumed to have been lifted to their positions as the result of the volcano's elevatory force. A somewhat similar or identical relation is presented by the Puy Chopine, in the Auvergne, where, as we are informed by Scrope and others,

---

* It is but proper to add that several geologists have suggested, in conversation with the author, the broad possibility of the Pelée tower having had this origin or pointed out the difficulties that lay in the way of accepting the more general view.

† Geologische Forschungen in den Kaukasischen Ländern,'' ii., pp. 67–78.

great blocks of elevated granite, sandwiched between trachyte, and constituting a portion of the basal rock of the volcano, now form part of the upper moiety of the dome and point unequivocally to elevation at a time or times of eruption. Other examples of this kind in the past histories of volcanoes could be cited, and, doubtless, many more than are at present known will be found when the craters of volcanoes, active and non-active, will have been more accurately investigated than has been the case until now.

The question of depth to which the Pelée tower descended within the throat of the volcano, assuming it to have been an ancient core, cannot be profitably discussed. It can merely be stated on this hypothesis that the accretions to height which followed every summit disruption and abasement were merely the expression of a further portion of the core thrust out.*

Did one need any direct evidence to support the view that I have set forth regarding the structure of the Pelée tower, it could easily be found, it seems to me, in the condition of parallel activity which the volcano has all along maintained at the summit,—namely, the construction of a fluidal dome (cone) and the simultaneous erection of a rigid spine or tower. This divergent condition is hardly explicable on the theory of the almost instantaneous cooling of the outwelling lava, whereas it it entirely consonant (and only what one should expect to find) with the notion of an ancient upthrow. Even as late as the beginning of the present year (January 3, 1904), what is de-

---

* For individual views on the structure and nature of the Pelée tower see: Israel C. Russell, " The Pelée Obelisk," Science, Dec. 18, 1903; Jaggar, " The Initial Stages of the Spine on Pelée," Amer. Journ. Science, Jan., 1904; Prof. N. H. Winchell, Amer. Geologist, 1904. Also, Branner, on the " Peak of Fernando do Noronha," Amer. Journ. Science, Dec., 1903. In a paper on the " Criteria Relating to Massive Solid Volcanic Eruptions" (Amer. Journ. Science, April, 1904), Prof. Russell cites a number of instances from among the American volcanic fields—Panum Crater, in the Mono Lake region, California; the tower-rock of the Bogoslov eruption of 1883; Pauline Lake Crater, Oregon—where structures thought to be analogous to the Pelée tower have been developed. These are all explained on the hypothesis of a rapidly solidifying viscous lava, thrust out in the manner that has generally been assumed for the Martinique tower; but to whatever extent these may share the Pelée type of structure, it seems to me that they receive an at least as acceptable interpretation in assuming that they are merely extruded ancient cores (necks). One may reasonably hold that such extruded cores must exist somewhere, and it seems to me that careful search will reveal many among structures which have hitherto received wholly erroneous interpretations.

scribed as being the remains of the ancient needle was reported by the French Commission to be rapidly rising, while the dome remained stationary or was being lowered through disruptions and cavings; and on November 11, 1903, a needle 15 metres in height, which stood on the western side of the dome, was reported to have disappeared. In the illustration that appears on Plate VII*a*, from a photograph taken in the month of March of the present year (1904), the jagged stock of a new obelisk or tower, unless it be the basal portion of the original tower that was destroyed, is seen overtopping the true summit of the mountain, and rising from its central supporting dome.

IV

THE chief features of the Pelée eruptions and their attending phenomena are discussed in detail in my "Mont Pelée and the Tragedy of Martinique," and such new observations as have been made only tend to emphasize the extraordinary nature of these eruptions. Pelée, indeed, stands out unique among all the volcanoes of the globe, and the object lesson taught by it is the most impressive and perhaps most important that appears in the records of vulcanology. The principal features and effects of its activity may be paragraphically summarized as follows:

A. A disturbance in the electro-magnetic field of our planet which in magnitude surpassed all hitherto recorded disturbances of this nature, the almost immediate and consentaneous effects being registered at the widely removed magnetic observatories of Cheltenham (in Maryland), Baldwin (Kansas), Toronto, Stoneyhurst, Val Joyeux (France), Paris, Potsdam, Pola, Athens, Honolulu, Zi-ka-Wei, in China, and elsewhere, the traverse of the disturbance being in all cases about two minutes of time. No previous volcanic eruption, not even the paroxysmal destruction of Krakatao in 1883, is known to have produced any magnetic disturbance other than of a local character.

B. The production of electric or pyro-electric illuminations in the volcanic cloud seemingly far surpassing those that had ever before been noted, and presenting features that had not hitherto been recorded.

C. The propagation of sound-waves to distances of 800 (and probably 1000 or more) miles, the explosion of May 8 having been heard with terrific intensity at Maracaibo, the sound, as likewise that accompanying the eruption of August 30, seeming to come from above.

D. The transmission of a shock-wave, or earth tremor, as would appear from the single observation made at Zi-ka-Wei, in China, passing completely through the earth,—a condition that had only once before been noted (in connection with the Krakatao eruption).

E. The formation of a remarkable series of "after-glows," or brilliant red skies, which doubtless made the passage over the entire earth, and were observed off the Venezuelan coast, at Los Angeles

38

(California), Honolulu, Bombay, Funchal (Madeira), in most parts of Europe, from Italy to England, and along nearly the entire Atlantic border and the central portion of the United States. These skies, with the attendant Bishop's ring, were less brilliant than those which followed the Krakatao eruption, occupied a position much nearer to the earth's surface, and travelled with somewhat less than half the velocity.

F. The emission of prodigious quantities of steam and ash, the steam-column passing at times vertically through the zones of both the trade and anti-trade winds, and to heights above the summit of the volcano estimated to be from four to six miles. The furthest distance at which the falling ash was noted on the surface of the sea appears to have been about 700 (900?) miles.

G. The issuance of an explosive tornadic blast, of a nature perhaps not yet entirely understood, whose death-dealing and destroying effects have no other event in the earth's history to compare with it. The event of August 30 was a repetition of that of May 8.

H. The extrusion from the crater-summit of the volcano of a giant core of solid lava, a veritable tower or obelisk, which at its most lofty period (May 31, 1903) rose to about 1020 feet, with a thickness at the base of 350–500 feet (shortest and longest diameter).

I. The eruptions of Pelée took place in times of atmospheric stability, were unaccompanied by earthquake movements, and had no relation to distinctive phases either of the moon or of the sun.

J. The ejected products, exclusive of the tower, were of a fragmental,* aqueous, and gaseous nature, there having been no true lava-flows (at least, not beyond the crater-limits).

---

* The fragmental products of the Pelée eruptions are essentially a highly acidic hypersthene-andesite, whose general composition, as determined by the analyses of Hildebrand, Mirville, Pollard, and others, may be stated to be SiO_2, 53–62 per cent.; Al_2O_3 (and Fe_2O_3), 20–30 per cent.; CaO, 6–10 per cent.; MgO, 2–4 per cent., $+ N_2O, K_2O$, and H_2O. This does not differ essentially in composition from the andesitic rocks which form the old stock of the volcano, and which are so largely distributed over the island of Martinique. A true cyclic succession of the volcanic rocks, following the Richthofen view, would seem not to have been here realized, although it is true that a large part of the ejected material was in all likelihood from the old stock of the volcano. Giraud, from a determination of a few fossil remains found in the tuffs of Trinité and Marin (*Turritella tornata*, also found in the Miocene of Panama—*Pecten (Amussium) subpleuronectes*, and *Aturia aturi*) and on other grounds, assumes the earlier volcanic outflows to have begun in the Oligocene, and to have been carried through the Miocene period (Bull. Soc. Géol. de France, Nov. 17, 1902, p. 395; *ibid.*, Feb. 16, 1903, p. 130).

K. No marked alteration in the coast-line or in the height of the land has thus far been noted in the region of the disturbance. There is, on the other hand, strong reason to believe that violent disturbances took place along the oceanic floor near by, even if not necessarily disturbing in marked degree the position of that floor.

L. Each violent eruption was accompanied by a vertical displacement, of short duration and with infrequent oscillations, of the sea-level, the surface rising on both the east and the west side of the island about three feet.

M. There can be no doubt as to a chorologic relationship existing between the activities of Pelée and the Soufrière of St. Vincent.

The following additional notes and observations, bearing upon the different topics indicated in the several paragraphs, are given towards further completing the scientific history of Martinique's remarkable volcano.

(*a*) The magnetic observations made in different parts of the world, and collected by the Coast and Geodetic Survey of the United States, show that the remarkable disturbances in the magnetic field to which reference has been made had a common initial time over the entire globe,—namely, 7h. 54.1m. A.M., Saint-Pierre local mean time. The data were obtained from observations made at twenty-six observatories encircling the globe.*

There is hardly room to doubt that the transmission of the disturbance was effected through the heart of the earth, and did not follow a surface course. The passage of an electro-magnetic or electric current through the earth opens out an interesting inquiry as to possible effects that may have been produced by it. Some of these effects could, perhaps, be held to be productive of a certain form of volcanic energy in distant regions, or at least to be an inciting force.

(*b*) In my account of the extraordinary pyro-electric display seen in the volcanic cloud of the evening of August 30, the night of the destruction of Morne Rouge and other settlements, I referred to the peculiar figures which, with extreme electric brilliancy, moved and

---

* L. A. Bauer, "Magnetic Disturbances during the Eruption of Mont Pelée on May 8, 1902." Paper read before the International Geographic Congress, Washington, September, 1904. Dr. Baur also calls attention to the interesting fact that a similar magnetic disturbance was noted on April 17–19 (also on April 10), "covering the period of the Guatemalan earthquake" (Quezaltenango); this is, again, precisely the period when Pelée was first significantly active.

flashed directly overhead,—straight and serpentine lines, single and in parallel series, tailed and tailless circles, rocket-stars, etc. It seems that most of these figures had nowhere been noted before the May eruption, although something analogous had been observed in New Zealand at the time of the Tarawera eruption of 1886. A note communicated by Mr. Powell, Curator of the St. Vincent Botanic Gardens,* on the great eruption of the Soufrière of September 3, 1902, refers to serpent electric flashes in the sky at that time. Doubtless, these were of identical nature with those observed in the Pelée cloud, although no reference is made to the lines occurring in parallel associations. I find, however, as illustrated in *La Nature* for January 30, 1904, and described by Em. Touches (*La Forme et la Structure de l'Éclair*) in a review of Prinz's recent work, that a form of undulating quintuple lightning was observed in Paris on July 29, 1900.

(*c*) It is interesting to note that, while the noise of the Pelée eruptions of May 8 and August 30, as noted at Maracaibo, 800 miles distant, at Carúpano, on the Venezuelan coast, and at Port of Spain, on the island of Trinidad, appear to have come from above, or, as stated by Consul Plummacher, to have originated in the clouds, such detonations have very generally been described as being subterranean, the propagation of the sound-waves being readily facilitated by the solid rock-masses. Thus, Humboldt, referring to the eruption of Cotopaxi in 1744,† states that the propagated noise, which was heard at a distance of at least 436 miles, was surely subterranean; and Scherzer, who received testimony of witnesses of the event, ‡ states that at the time of the great eruption of Coseguina the detonations, which were carried hundreds of miles, appeared subterranean. Probably no exact reason can be assigned for these differences in sound-carriage; some of the anomalies of this transmission have already been pointed out in my report.

(*e*) It has before been said that the red skies or "after-glows" which followed the Antillean eruptions made, with little doubt, a full traverse of the earth's atmosphere, the numerous and distant points at which these magnificent phenomena were observed giving sufficient evidence in support of this assumption. The fact that the Soufrière

---

* Journ. Geol. Soc. London, Feb. 10, 1903.

† Cosmos, Bohn edition, i., p. 203.

‡ Wanderungen, 1857, pp. 479 et seq.

THE TOWER OF PELÉE

and the volcano of Santa Maria, in Guatemala, were also in eruption during the period of Pelée's activity, and throwing out vast quantities of ash, has naturally made it impossible to correlate the afterglows, especially those of the later dates, with the individual eruptions. I have myself noted, around Philadelphia and New York, the most brilliant sky-glows, unquestionably of the volcanic type, at recurrent periods in December, 1902, and in late January, 1903. Mr. Backhouse * notes their occurrence in association with a solar corona (Bishop's ring) at Sunderland, England, at the end of June, on October 30, November 1 (at its full magnificence), 17 and 18; at Torquay, on November 6 and 10; and at Dundee, on December 1. M. Enginitis, the Director of the Athens Observatory, notes the after-glows of October 25 and later, beginning a few minutes after sunset, and rising, like the glow from a conflagration, to a height of 45 degrees. A similar glow is stated to have followed the eruptions of Etna in 1831.† More recently Professor Forel has described ‡ the Bishop's ring carefully studied by him at Morges, on Lake Geneva, the identity of which with the ring observed by Bishop in Honolulu, in 1883, is stated to be absolute. The period of visibility of the new ring appeared to have extended from August, 1902, to December, 1903, § and seemingly was a continuous one from favorable points of observation. Forel noted it (practically) every day when he was placed in positions removed beyond the dust-zone of the lowland plains,—from the Rochers de Naye, in Valais, from Pilatus, St. Gotthard, the middle and upper slopes of Mont Blanc (where it was also observed [Montanvers] by Laurence Rotch on August 20, 1902). ‖ Other points of observation recorded are: Arnsberg (November 19, 1902, March 21–22, 1903, by Dr. Busch), Heidelberg (January, 1903, by Prof. Max Wolf), Zürich (January, March 27–28, July 7, 8, 9, and later of the same month, 1903, by Dr. Maurer), St. Petersburg (October 5, November 9, 1902, January 21, February 10, 18, 23, March 17, April 5, May 29, July 26, 1903, by Rykatcheff), Lucerne (July 26, 1903, by Dr. Arnold), Frankenfeld, Clarens, Hoh-Königsburg, in Alsace, etc. Forel concludes that the

* Nature, Dec. 25, 1902, p. 174.
† Communication to the Academy of Sciences of Paris, Revue Scient., Dec. 20, 1902, p. 787.
‡ Archives des Sciences physiques et naturelles, Oct. 15, 1903; Feb. 15, 1904.
§ Helm Clayton, in Nature, Jan. 21, 1904.
‖ Nature, Oct. 29, 1903.

42

virtual continuity in appearance of the Bishop's ring is, at least, presumptive proof that the ash-belt in the *higher* regions of the atmosphere was a continuous one. This is thought not to have been the case with the *lower* belt or zone of ashes which originated the crepuscular glows, for these appeared only at irregular intervals, differing in this respect from the glows following the Krakatao eruption.*

A most interesting observation has latterly been made touching the distribution and retention of the volcanic particles in the atmosphere,—namely, that they have served as a cushion or screen to reduce the intensity of solar radiation. According to Henri Dufour,† such a diminution of radiant measure was noted, among other places, at Clarens, Lausanne, Heidelberg, Warsaw, Washington, etc., beginning in December, 1902, and continuing but steadily diminishing to March, 1903. This opacity of the atmosphere, which is attributed to the Antillean outpourings, and may be due directly to easier condensation of vapor under the influence of ash-nuclei, is evidenced: 1, by diminution in the intensity of the solar radiation; 2, diminution of the optical transparency of the atmosphere; 3, diminution in the sky's polarization; and 4, displacement of the neutral point of Arago and Babinet. These several conditions had been clearly noted in the atmospheric disturbances following the eruption of Krakatao. Ladislas Gorczynski, who has been following up Dufour's observations and is inclined to accept Dufour's interpretation of the phenomena, notes that the diminution was observed in Warsaw, Poland, as early as May, 1902, and that from that time it increased progressively until the spring of 1903; it had practically disappeared before the close of that year.‡

(*g*) There is little to add to the views that I have already expressed with regard to the nature of the tornadic blasts which brought about the appalling catastrophes of May 8 and August 30. Most observers appear now to be pretty well agreed that the main engine of destruction was steam in a superheated condition or in a condition of high tension and extreme temperature. In how far the work of death may have been assisted by the association with this tornadic

* Archives des Sci. P. et N., Feb. 15, 1903.

† Comptes-Rendus, cxxxvi. pp. 713–715; Arch. des Sci. Phys. et Nat., Oct. 15, 1903, pp. 459, 460.

‡ Sur la Diminution de l'Intensité du Rayonnement solaire en 1902 et 1903, Comptes-Rendus, Feb. 1, 1904, cxxxvi. p. 255.

steam of poisonous, not necessarily inflamed, gases, perhaps will never be known. In my first report * I was inclined to attach first importance to the effects of one of the heavier carbon gases, it having appeared to me at that time that the steam played the less important part so far as the extinction of human life was concerned. After my later visit to the island, when I was made a close observer of, if not an absolute participant in, the second death-dealing eruption, and had the opportunity of almost immediately studying the effects of this eruption, I found it necessary to modify my views with regard to the destroying agent, and to attribute the major work of destruction to explosive steam. The participation in this work of inflamed gases was nowhere apparent; indeed, the evidence obtained from the character of the vegetation, from unburnt wood-work, from the unaffected clothing, and from the experiences and sensations of the wounded and dying who went through the storm, showed that there could have been no such participation. At the same time, the absolute annihilation wrought has always appeared to me a puzzling feature in the work of steam alone, and while it may be admitted that the overthrow of Saint-Pierre was due virtually to this one cause, or to the tornadic work which it impelled, it seems not unlikely that the work of human destruction was bound in with accessory conditions, some of which may never be known to us. With regard to the possible assistance of asphyxiating gases, and as bearing upon my first-expressed view, it is interesting to note that M. Moissan, who has made a close study of the fumarole gases of Pelée, finds the quantity of carbon oxyd so large as to warrant the assumption that it must have been present in sufficient measure in the exploding cloud of the main eruption to have caused, through toxic inhalation, the deaths of at least a large portion of the populace. Other gases found were hydrogen, methane, and argon, the last two also found among the gases of the waters of Luchon (addressed to the *Académie des Sciences* of Paris, December 15, 1902; *Revue Scientifique*, January 8, 1903).*

---

* Published in McClure's Magazine for August, 1902.

† Moissan has since found the carbon oxyd gas in large proportion among the fumarole products of the Soufrière of Guadeloupe. It is interesting to note that Boussingault, from observations made upon the volcanoes Tolima, Quindiú, Puracé, Pasto, Tuqueres, and Cumbal, of the Equatorial Andes, found that their chief gaseous emanations were water-vapor and carbonic acid, the sulphurous acid present being considered accidental; and it is remarked that even where the odor of sulphur is

I have always felt that electric discharges were also largely responsible for the destruction of life; indeed, the case could hardly have been otherwise, for, as we are informed by competent witnesses, the death-dealing cloud was charged with electricity, short flashes passing at rapid intervals from point to point. This same feature had also been observed in the descending cloud of June 6, and Flett and Anderson refer to it in their description of the cloud of July 9, 1902. During my latest visit to Martinique I was informed, by one who was saved from the destruction of Ajoupa-Bouillon (although losing his family in that terrible disaster of August 30), that the descending cloud that wrought the havoc was flashing with electric lines and sparks, resembling artificial fireworks.

I have elsewhere expressed my view that the destruction of Pompeii was in all probability caused by a volcanic discharge similar to that which brought about the annihilation of Saint-Pierre, and that the phenomena of the Vesuvian eruption of the year 79 and of Pelée were largely similar. It seems not unlikely that there may have been eruptions from other volcanoes, the conditions of which have not been properly investigated, which had much in common with what is assumed to be the distinctive features of the Pelée explosion. Thus, it is noted by M. Fouqué, in his work on Santorin, that at the time of the eruption of the year 1650 the dead bodies of a number of sailors were found on a drifting vessel several miles from the seat of the eruption, and exhibiting abdominal and head inflation, protruding tongues, and inflamed eyes. These features of corporeal distortion were a marked characteristic of the killed in both the Pelée eruptions, and have been attributed to special conditions surrounding the death-stroke. It is interesting to note that Dr. von Volpi, describing his own personal observations on the great eruption of Vesuvius in April, 1872, refers to the terrific scalding that was brought about by superheated steam, and the resulting red scars on the human flesh: *"Man bringt einen Verwundeten, dessen Haut und Fleisch verbrannt sind und krebsroth aussehen. . . . Die Verwundungen rührten nicht etwa von Berührung der feurigen Lava her, sondern von dem glühendheiszen Dampfe, der von ihr*

---

strongly felt the actual quantity of the gas present is very small when compared with carbonic acid, which is maintained even at the high temperature of 334° C. (Annales de Chimie et de Physique, vol. lii. p. 23, 1833.) Bunsen also found that carbonic acid vastly preponderated among the gaseous exhalations of the Iceland volcanoes.

*ausging und bei einer Hitze von 800 Grad alles versengte und ver-
brannte, was in seiner Nähe war."* *

Drs. Flett and Tempest Anderson, in their report to the Royal
Society of their investigations in Martinique, assume that the black
tornadic cloud which wrought or assisted in the destruction of Saint-
Pierre made its phenomenally swift descent from the crater purely as
the result of the acting force of gravity. It seems to me that, had these
investigators visited the basin of the Étang Sec and studied its characteristics from near, they could not have arrived at this conclusion,
for no form of discharge originating within the basin itself, unless it
had been directed in its initial movement laterally, could have carried
the materials of disruption much, if at all, beyond the boundaries of
this basin. The capacity of the Étang Sec was at that early period
sufficient to hold a far greater accumulation than that which swept
down the volcano's slope. And even assuming the possible transfer-
rence of the ejected material beyond the crater borders, it is not conceivable how such material, whether cushioned with steam or not, could,
with so low an angle of slope and with the barring obstructions in the
path of passage (among others the ravine of the Rivière Seiche), have
acquired that prodigious velocity which all witnesses of the eruption
agree in saying that the destroying blast had. The linear distance
from the basin of the Étang Sec, the floor of which at its opening occupied a position of about 2600 feet above the level of the sea, to Saint-
Pierre was about four miles, so that the average slope of the mountain
was almost exactly 1 in 8; and yet it is certain that the destroying
cloud swept down at a rate of at least $1\frac{1}{4}$ miles, and not unlikely $1\frac{1}{2}$
miles, per minute, a rate of descent but little greater than that which
had been observed in several of the more recent eruptions. My own
studies in the field leave no doubt in my mind that this so-called "avalanching" of the cloud and its contained material could only have been
produced as the result of a lateral or descending explosive shock, a
conclusion which has also been reached by Lacroix and Giraud and
which is amply confirmed by the numerous later discharges that have
been observed and studied by different investigators since the May
eruption. On June 5, 1902, I was myself witness to one of these lateral
eruptions, still in the early period of the volcano's activity, when the
discharge for a considerable distance was carried through the air with-

* Unsere Zeit, Leipzig, 1872, p. 397.

out at all touching the slope of the volcano. Professor Lacroix, in his reports to the French Academy, refers to a number of discharges of the *nuée ardente* breaking out laterally from the base of the obelisk surmounting the crater-cone, and taking almost invariably a course down the sedimented valley of the Rivière Blanche,—the course that was followed by nearly all the eruptive clouds of the volcano. The cloud of September 9, 1903, made its way to the sea in five minutes. But no further proof of the explosive discharge of the black cloud need be had than the observation made by the French Commission,* that most of the *nuages denses* at about this period ascended vertically.

In discussing the nature of this lateral blast, which in a way is perhaps comparable to that which in 1888 dislodged a quarter of the summit of Bandai-San, I have referred to it as an explosion "in free air under a heavily depressing cushion of ascending steam and ash, and with surrounding walls of rock on three sides and more, to form an inner casing to nature's giant mortar." † Knowing, as we now do, that probably at that time the chimney of the volcano was already at least partially plugged, it is made easier to assume the deflection of the tornadic discharge through cushioning.

(*i*) It has been remarked as one of the peculiarities of so violent an eruption as that of Pelée on May 8, and equally so on May 20 and August 30, that there should have been no free flow of lava, a condition that was long ago indicated by Leopold von Buch in the case of the active volcanoes of the Andes generally. Neither in the great eruption of the Soufrière of 1812 nor in the more recent eruptions of that remarkable volcano has there been any lava-flow, that which has been described as lava in 1812 being merely a detrital and mud discharge similar to the discharge of 1902. Some geologists have attempted to measure the explosive force of different eruptions by assuming the quantity of ejected lava as the determinant of this force,—the greater the amount of lava emitted, the greater the force of the volcano. It would seem, however, that the opposite conclusion would more nearly represent the truth, for we find that nearly all the great paroxysmal discharges were unaccompanied by lava-flows, or at least by lava-flows of any magnitude. Such was the case, for example, with the eruptions of Galunggung, in Java, of Temboro, or Sumbawa, in 1815, of the

---

* La Colonie, Sept. 15, 16, 1903.

† "Mont Pelée and the Tragedy of Martinique," p. 317.

Soufrière, in 1812, and the eruption of later date, of Coseguina, in Nicaragua, in 1835, of Krakatao in 1883, and the eruption of Pelée of 1902. To these examples, and many others that might be cited, should be added the first recorded and historic outburst of Vesuvius in the year 79, a condition in marked contrast to subsequent less paroxysmal eruptions and in which the flow of lava took a conspicuous part. It might be assumed, in explanation of this seeming inversion of force and effect, that in the paroxysmal types of eruption the quantity and force of the pent-up steam are more than sufficient simply to lift lava, but also blow it to pieces, and produce those enormous volumes of ejected material which have buried or overthrown towns and villages, and otherwise defaced the landscape over vast distances. Volcanoes of a less paroxysmal type will pour or "well" out the lava in quiet streams, not necessarily accompanied by any marked form of explosive action.

(*l*) The singular oscillation of the sea-surface, the rise of the water by about three feet, which was noted as an accompaniment to several of the more forcible eruptions of Pelée, on both the west and east coasts (Fort-de-France, Trinité, etc.),—a phenomenon which I myself witnessed on the morning of June 6,—may possibly have been due to direct volcanic shocks impacted upon the floor of the ocean. The existence of oceanic disturbances off the west coast of Martinique at the time of, or preceding, the great eruptions of Pelée can hardly be doubted. The successive breakages of the different cables precedent to the great eruptions, and other facts connected with the attempted location of the disrupted ends of the cables, prove this condition almost beyond a doubt. Krebs has latterly called attention, in a paper on tidal fluctuations as related to volcanic phenomena,* to an extended marine disturbance which traversed the entire length of the Guatemalan coast on the 16th and 17th of April (1902), one and two days in advance of the great earthquake which wrecked a portion of the town of Quezaltenango, and which was almost coincident with the first breaking into activity of Pelée. A similar oceanic disturbance was noted on May 4, the day in advance of the mud discharge from Pelée which overwhelmed the Usine Guérin.

---

* Globus, July 30, 1903.

V

SOME THOUGHTS ON VOLCANIC PHENOMENA SUGGESTED BY THE ANTILLEAN
ERUPTIONS

THE broad territory in the Caribbean-Gulf region which was covered by the seismic and volcanic disturbances of 1902 is very noteworthy. From southern Mexico in the west to the Lesser Antilles in the east we have an interval in a direct line of not less than 1800 miles, and along or near this line disturbances have been registered in Costa Rica, Nicaragua, Salvador, Guatemala, and Mexico. The remarkable crowding of the phenomena is such that one cannot well resist the conclusion that they are all interrelated or hold a mutual relation to a single inciting cause, and are not coincidental in their occurrence. Thus, as Rockstro and Sapper have also directed attention, the earthquake of Quezaltenango, in Guatemala, took place on almost the exact day, April 17–18, on which Pelée first seriously manifested its new activity; and the volcano of Izalco, in Salvador, ordinarily one of the most active of American volcanoes, but whose eruptive energies had calmed down for a number of years, started upon a new period of eruptivity almost immediately after the earthquake of April 18 (the effects of which were noted from Salvador and Honduras to Mexico), and was in energetic action on May 10, two days after Pelée's great paroxysm.* It was not without reason, therefore, that Milne advanced the view that the earthquake of Quezaltenango was the real initiator or instigator of the disturbances that followed rapidly upon it: it took the lid off the boiling pot, and the pot exploded. But one may reasonably extend the history of the disturbances so as to include the earlier earthquake which on January 14 (16?) in great part wrecked the town of Chilpancingo, and the later eruption (beginning on October 24, 1902, and continuing to November 15) of the volcano of Santa Maria, in Guatemala,—a volcano situated close to the field of Quezaltenango, and of which no antecedent eruption has been chronicled for ante-Columbian

---

* Rockstro, Nature, Jan. 22, 1902; Sapper, Centralblatt für Mineralogie und Geologie, April, 1903.

49

or historic times,*—together with the reawakening in February and March of Colima in southern Mexico.†

Assuming, as we do, the continuity of these various forms of disturbance in the Caribbean region, we are necessarily driven to the conclusion that the inciting force of the disturbances was regional in its extent, and not local, and was in no way concerned with localized rifts and subvolcanic fissures, the penetration of sea-water, of land-water, etc. The inquiry is, What was the nature of this inciting force? What are the particular conditions in our planet which at varying intervals bring the products of eruptive action (steam and lava) to the surface or initiate volcanic phenomena?

On the generally accepted hypothesis that layers, beds, or pockets of molten rock-material lie within the earth at no very great distance beneath the surface,—perhaps 20 or 30 miles, or even considerably less,—or that potentially molten rock occupies this position and on release of pressure would assume the fluidal condition, it is easy to postulate that the application of strong vertical mechanical pressure, whether directed over local or over broad areas, might "squeeze" this material to the surface,—force it into channels where an exit is made possible. Such a form of pressure might be, and almost certainly is, furnished by the weight of subsiding areas of the earth's superficial zone or crust, and it is hardly a coincidence that all active volcanoes (and, inferentially, this may be made applicable to all extinct and ancient centres of eruption) are placed relatively to the land-masses of the globe in regions of marked instability or weakness and where the exercise of pressure produces work. Their absence from the major areas of the superficial crust where ages of construction and strain have established a rigid stability proves that the dynamic force underlying volcanic action is made operative only where a way, whether by forcing or dislocation, has been prepared for it.

I have elsewhere ‡ stated my belief that a subsidence of the floor of the Caribbean Basin, causing displacements of equilibrium and forcing molten and other material to the surface, was the inciting force of the Antillean eruption; and it still appears to me that his hypothesis alone satisfies the conditions which the broadly distributed phenomena call

---

* Rockstro, Nature, Jan. 22, 1902.

† Ezequiel Ordonez, Les Dernières Éruptions du Volcan de Colima, Mexico, 1903.

‡ Mont Pelée and the Tragedy of Martinique: "The Volcanic Relations of the Caribbean Basin."

50

for. Unfortunately, in problems of this class, where the more important facts or conditions are destined to remain obscure, it is impossible to absolutely unite cause and effect, and it is only a strong probability that can be assumed.* The coincidental association between the distribution of volcanoes and the larger mountain tracts, and the fact that both lie close to the sea-board, do away entirely with the necessity of invoking the acting presence of sea-water as a *condition* of vulcanism. Volcanoes, like mountains, lie close to the sea-border because the super-crust is there weak, and *not* because the ocean-water is needed in their making. There are no active volcanoes in or adjacent to extensive strips of modern flat land along the sea-board or elsewhere where these are free of mountain elevations; nor are there active volcanoes associated with ancient mountain masses (as in Norway, Sweden, Greenland, Labrador, Brazil), unless these mountain masses, whether on the sea-board or not, have lying in with them other mountain parts of much newer date.

It might be expected, in proof of a subsiding region, that somewhere along its contours evidence would be found of a rise or "squeezing up" of the adjoining land-mass, whether in the form of a bodily uplift or of a plication or series of folds. As pertaining to this particular region, it is interesting to note that so far back as 1890 Prof. Shaler, in a paper on "The Topography of Florida," had expressed the view that this singular projection of the Atlantic contour of the United States was due to a squeeze between the subsiding areas, the Gulf basin on one side and the Atlantic basin on the other; † and I have held to the same view both as regards the peninsula of Florida and the opposed peninsula of Yucatan.‡ The subsidence of the Gulf basin has always seemed to me the (main) force that uplifted the eastern and central sierras of Mexico, whose buttresses constitute the inner core of the great central plateau. The evidences of comparatively recent uplift in the Antillean tract are seen on the terraces or oceanic beaches of Cuba, Jamaica, etc.; and more recently they have again been carefully studied by Spencer and Sapper in the Lesser Antilles, where nearly all

---

* See an early paper by Starkie Gardner on the correlation of volcanic eruptions and oceanic pressure or subsidences, in the Geological Magazine, June–July, 1881.

† Bulletin Mus. Compar. Zoölogy, xvi.

‡ "Geological Researches in Yucatan," in Proc. Acad. Nat. Sciences, Phila., 1891, pp. 136–158.

the islands bear the marks of this uplifting. On St. Lucia the coral strands, as found by Sapper, rise to 40–150 metres; on Dominica, to 15 and 60 metres.* On Martinique the old sea-washed surface is clearly distinguishable. In just what manner this late elevation of the islands was effected cannot positively be told, but it suggests lateral thrust similar to that which has been assumed for the peninsulas of Florida and Yucatan, the acting force being the Atlantic basin on one side and the Caribbean on the other. This view has already been in a way expressed by Michel Lévy, in his paper on the Antillean eruptions, published in the *Revue Générale des Sciences Naturelles*, June, 1902, and it is one in kind which is generally accepted in explanation of closely corresponding phenomena of the Mediterranean region of Eurafrica.

We may assume three resultants of any applied telluric pressure: 1, the forcing out direct of a pocket of molten magma (perhaps exemplified in some of the quiet outwellings of lava, which flow without explosion); 2, releasing from pressure parts of the interior which are at the critical temperature of melting, and permitting them to be converted into lava, with subsequent extrusion; and, 3, the superheating of water-containing rocks of a higher horizon by forcing to them the heated rocks of the deeper interior, and by this contact producing explosive force. This last condition is accepted by Stanislas Meunier as the basal explanation of volcanic phenomena generally,† and it can hardly be doubted that it sufficiently explains many of the explosive forms of eruption.‡

We may, following Mallett, establish calculated temperatures in the interior as the result of earth-pressure, and by other means assume the equivalent of mechanical work that would be furnished; but it does

---

* Centralblatt für Mineralogie, Nos. 9, 10, and 11, 1903.

† Acad. des Sciences meeting, Jan. 12–19; Revue Scientifique, Jan. 24, 1903, p. 120.

‡ A somewhat similar conclusion, bearing upon the Antillean eruptions, is expressed by Prof. Robert T. Hill, in his report published in the National Geographic Magazine (1902, xiii. p. 266), where he says: " The synchronism of this eruption (Pelée) with that of St. Vincent, a hundred miles distant, and volcanoes of similar andesitic character in Central America, to say nothing of disturbances reported in volcanic areas throughout the world, is strangely, almost positively, suggestive that the cause of the eruption of Pelée was not the development of a local fissure suddenly letting the water of the sea down to the depths of the hot magma, but, upon the contrary, resulted from a widely occurring disturbance within the interior of the earth's magma, which caused it to rise to meet the upper wet zone, rather than the water of the latter to descend to it. and which is as yet inexplicable."

52

not seem to me that such calculations at this time afford us more than interesting conjectural results, the verity of which, in any application to existing conditions, may be very wide of the real truth. Calculations of this kind, while they undoubtedly have their value in forcing other comparisons with them, have very generally proved to be suggestive rather than fundamental, and they rarely furnish data that are substantial and resist attack. Similarly, any discussion of the condition in which a molten (or potentially molten) magma lies within the earth, however interesting and fruitful of generalization it may prove, can hardly afford more than conjectural results. Stübel * has ingeniously attempted to show the succession of volcanic events in the earth's history, but they are wholly impossible of demonstration, either as to the successional building up of the earth's crust or of the two forms of volcanoes which he designates monogenetic and polygenetic. One may safely assume the existence of such magma without feeling that the condition of its occurrence or formation is necessarily a party to the problem under consideration.

The most *catastrophic* events of vulcanism have almost invariably been developed by island volcanoes or by volcanoes that are absolutely situated on the sea-board. Among the examples of the paroxysmal types of eruption illustrating this view may be cited the explosions of Papandayang, in Java, in 1772; of Asamayama, in Japan, in 1783; of the Soufrière of St. Vincent, in 1812; of Temboro, in Sumbawa, in 1815; of Coseguina, in Nicaragua, on the Bay of Fonseca, in 1835; of Santorin, in 1866; of Krakatao, in 1883; of Tarawera, in New Zealand, in 1886; of Bandai-San, in Japan, in 1888; and of the Soufrière and Mont Pelée, on the islands of St. Vincent and Martinique, in 1902. To this same class must be added the earliest historically recorded eruption of Vesuvius, that of the year 79, when Pompeii, Herculaneum, and other fair towns and villages in the region of or about Campania were overwhelmed.† The existence of the islands themselves, or the severance of parts of land from a united mass, may in most cases be taken as evidence indicating strong crustal movements and corresponding

---

* Ein Wort über den Sitz der Vulkanischen Kräfte, 1901.

† The date of the destruction of Pompeii is usually given as the 24th of August; but there can hardly be a doubt, if Pliny's relation is to be relied upon, that this event took place on the day following (25th),—the day on which the great black cloud is described by Pliny as having swept down the volcano's slope, veiling the landscape in darkness.

crustal debility. The lofty active cones of the equatorial Andes have had in the period of our knowledge regarding them no eruptions that were in any way comparable with these, and this is also true of the active cones of main and peninsular Alaska (Iliamna, Wrangel, Makushin, Sheshaldin), of Kamtchatka, of Mexico (Colima, Ceboruco, Jorullo (the facts connected with the eruption of the last seem to have been given in an exaggerated form to Alexander von Humboldt), Popocatepetl, and Orizaba).

The association between volcanic activity and breakage zones or lines has recently been well emphasized by Hoernes in his review of the recent eruptions,* in which the Antillean volcanoes, paralleled with those on the southwest side of the Apennines and of Hungary, are regarded as being placed over the inner breakage areas of crescentic mountain folds. Voltz, in a paper recently published on the disposition of the Sumatran volcanoes, has shown that all the volcanoes of that island are located on fracture-areas (*"Bruch-Zonen," "Bruch-Kessel"*), and pointed out the interesting fact that these fracture-areas are wanting in the non-volcanic portions of the island.† And Hauthal, in his memoir on the volcanic regions of Chile and Argentina, has emphatically pointed out, as opposed to the views expressed by Stübel and others, that most of the volcanoes proper of that region, as distinguished from the massive flows, distinctly conform in their alignment to the linear trend or main axis of the Cordilleras,‡ thus proving association with a long line of crustal dislocation or fracture.§

---

* Die Vulkanischen Ausbrüche auf den kleinen Antillen, Mitteil. naturwiss. Vereins für Steiermark, 1902, pp. xxxi. et seq.

† Die Anordnung der Vulcane auf Sumatra, Jahresb. d. Schles. Gesel. für vaterl. Cultur, July 24, 1901.

‡ Petermann's Mitteilungen, 1903, v.

§ While the foregoing was in press I received Part 3 of Geologischer Theil of K. Martin's "Reisen in den Molukken" (Leyden, 1903), dealing with Buru and the neighboring islands, in which the isostatic displacements, resulting in upheavals, lateral thrusts, etc., due to subsidence of blocks of the earth's crust, and the relation of these to volcanic phenomena are given full affirmative value. The views that I have expressed regarding the condition of the Caribbean Basin seem to have an absolute application in the Molucca Sea, where the fracture subsidence zones are pre-eminently the areas of volcanic disturbance and of local upthrusts (pp. 283–288). See also this author's earlier paper on the structure of a portion of the Caribbean Basin: "Reise nach Niederländisch West Indien," ii. Geologie, pp. 213 et seq , 1888 (the dismemberment of Curaçao, Aruba, and Bonaire).

The Source of Volcanic Steam and the Assumed Penetration of Sea-Water.—It may perhaps be at once admitted that we shall never be in a position to know to what extent oceanic water gains access to the earth's interior, in whatever way such access may be made possible. The almost uniformly fresh water that is obtained in artesian borings that have been drilled close to the oceanic border, to 1000 and 1500 feet or more, as at Atlantic City, Cape May, and elsewhere on the Atlantic coast of America, would seem to indicate that a zone of saturation, even in loose sands, gravels, and clays, is reached within very short limits. But this condition may be reversed along the floor of the deep sea, where the forcing strain is prodigious and wholly beyond comparison with what is exerted over the superficial zones. It would not be reasonable, therefore, to conclude too hurriedly,—even if the presence of sea-water is not a requisite in the causation of volcanic phenomena,— that under these conditions there is no penetration into and through the adjacent sea-walls or sea-floor, and this despite the enormous thickness of sedimentary and organic deposit that must have accumulated in the oceanic trough. To what extent such penetration may be checked by the expelling force of the heat of the earth's interior is, again, one of the questions to which geology may never give a final answer.

It has latterly been argued by Krebs,* that the evidence of recent explorations in desert regions does, as a matter of fact, indicate oceanic filtration to very considerable distances. Thus, it is claimed, on the basis of Natterer's report on *Chemisch-Geologische Forschungen,*† that sea-salt has been obtained from wells in the oasis of Siwah, in the Sahara (where nearly all the well-waters are saline), at a nearest distance of 160 miles (265 kilometres) from the coast; and at Bilma, at a distance of 840 miles (1400 kilometres), this obtained salt constitutes an important industry. Were it true, as was so long held, that the Sahara was merely a region of degradational sands, and bore no relation to any comparatively recent "Mediterranean" or oceanic sea, it would perhaps be difficult to account for the presence of these salines except on the theory of oceanic filtration. It is somewhat puzzling to understand why geologists should have so generally adopted this theory of the Sahara basin. The physiognomic character of the surface, espe-

---

* Flutschwankungen und die vulkanische Ereignisse in Mittelamerika, " Globus," July 30, 1903.

† Geographische Zeitschrift, Leipzig, 1899, pp. 190–209.

cially where it abuts against its northern "horizon," the Great Atlas and the Aurês mountains, should have sufficed to throw suspicion on this conception, especially as the evidence of the fossils obtained or described by Charles Martins, Pomel, Loriol, Coquand, Rolland, and others plainly indicated the sea-bottom at a time as late as the Cretaceous or the Tertiary period. But the oceanic character of the Sahara can now be said to be definitely demonstrated by the few fossils coming from it, to which Lapparent has latterly called attention, and which he has properly interpreted to be those of living specific forms. The great Saharan sea, which covered an enormous expanse in the north of Africa and which only at a very late geological period retreated from the continent, may thus sufficiently explain the presence of the saline deposits which have proved puzzling to some investigators.

It is no longer profitable to discuss the notion for a long time held, and still held in some quarters, that vast open fissures are from time to time formed along the oceanic trough, and that through these the oceanic waters suddenly find their way to assumed *loci* of volcanic activity, and there lend themselves to catastrophic ministration. No geologist has given satisfactory proof or evidence showing that such fissures ever had or could have been formed; or, assuming the possibility of their formation, that sea-water could have found its way through them to any great distance beneath the surface. The long-continued state of eruptivity of many volcanoes is in itself *prima-facie* evidence of the non-existence of sub-oceanic rifts, otherwise we should be obliged to assume for them a period of open life of incredibly long duration. The "fissure" was needed to accommodate the oceanic theory of volcanic phenomena, but it never had support from the facts of geology.

Assumed Penetration of Land-Water.—The relations of this problem are much more accessible than those of the last, and we are at once face to face with the common fact, as shown by springs and other waters, that the penetration of fresh waters is at least considerable. This penetration is not, however, directly through the substance of rock-strata, as the surface-inlay of rain-waters plainly proves, but through or along rock-rifts and fractures, bedding and sliding planes, fault-planes, etc.

If this penetration as is indicated by the evidence of temperature alone is not sufficient to satisfy the conditions of volcanic localization, it yet touches upon a possibility. The problem that presents itself in this connection, however, is not one of possibility or probability alone. It must be shown, or at least made probable, that, with all the conditions of penetration satisfied, the descending waters are in any way materially concerned in the phenomena that are so forcefully brought to the surface and in which the vapor of water plays so important a part.

It may be safely said that the facts of geology give no support to the view that sees an association between surface-water and volcanic phenomena. The activities of far-oceanic volcanoes, with their small catchment basins, wholly preclude the notion that their work is related to the amount of accumulation and descent of the local waters. The paroxysmal eruptions of Pelée and the Soufrière, 350 and 375 miles from the nearest point of continental land, and the sudden recurrences of paroxysms, with prodigious discharges of steam-water, after comparatively short intervals, are evidence in the same direction. The continuous activity of Stromboli, whose eruptions, in one form or another, have followed almost uninterruptedly for a period of 2000 years or more, is likewise opposed in its bearing to the view that has been held. It is true the argument might be advanced that the supply-basin of the surface waters need not necessarily be a near one; but this condition, in its application to the very large number of cases that it would be obliged to cover, is so conjectural and remotely probable that it need not be considered. It is a form of explanation that crops up only too frequently in geological discussion.

One of the forms of evidence that has been brought forward to support the view that the land-waters have much to do with the immediate causation of volcanic explosions is that, at times of eruption, it has frequently(?) been noted that the surface waters of the regions immediately disturbed underwent diminution in bulk, and in some instances entirely disappeared,—the land "drying up," as it were. That such conditions have taken place can hardly be doubted; the evidence of authority on this point is seemingly conclusive. It is not unnatural, therefore, to associate this disappearance with the phenomena that seemingly directly accompany it. But in very few of the cases that geologists have cited to prove this association are the data of so precise a nature as to permit us to state that the disturbance of the surface-waters was precedent, and not subsequent, to the actual eruption. That the latter condition was true in a number of cases is indisputable, and it is in no way surprising, as land-movements, whether true earthquakes or earth-tremors, are a frequent accompaniment of volcanic eruptions, and it would be but natural to find in such places displacements or even a complete obliteration of water-channels or basins. One might be tempted to conclude that the case could hardly be otherwise. On the other hand, in the remarkable paroxysmal eruptions of Pelée, which followed one another at exceedingly short intervals (May 8, May 20, June 6, July 9, August 30), there was practically no disturbance of the surface-waters, except of such which lay directly in the path of the erupted products; and, as is well known, the hydrant and fountain-waters were running from some of the Saint-Pierre spigots when the city was first explored after its destruction, on May 10, and continued flowing for ten or fifteen days. The belief that the waters of the Lac des Palmistes, the summit tarn, had suddenly been drawn into the volcano and were the cause of its first violent eruption, was purely fanciful, and was founded on the supposition that this lake, and not the Étang Sec, represented the true crateral basin of the volcano. There was no disturbance whatever in the basin of the Lac des Palmistes, beyond infilling with the discharge products of the several eruptions.*

It would not be difficult to cite numerous instances where volcanic eruptions have left unaffected the water-supply of the regions immediately about, whether by precedence or by subsequence; nor again to show where the force of the eruption had directly opened the way for a new distribution or even practical annihilation of either standing or running waters (Tarawera, Ilopango, etc.). Geologists have so far found it impossible to establish any relationship or concurrence between the periods of volcanic eruption and particular meteorological conditions of the atmosphere, whether of pressure or of rain-supply; and much less has it been possible to show that the intensity of eruption is in any way heightened by an excess of precipitation. Prof. Suess,† who has latterly attempted to show that the phenomena of certain hot springs (and geysers) are fundamentally volcanic in their nature, and are of a deep-seated, rather than of a superficial source, has well laid emphasis on the fact that these waters, in the quantity of their discharge or their periodicity, bear likewise no relation to the meteoro-

---

* It would appear from Rockstro's investigations that the water of the hot springs in the Santa Maria region of Guatemala was reduced in quantity after the earthquake of April, 1902. (Nature, Jan. 22, 1903.)

† Ueber heisse Quellen (Gesell. Deutscher Naturf. u. Ärzte. Verh., 1902.

logic conditions that surround them, and that their operations are performed alike in periods of drought and excessive precipitation.

The Hydrated Rocks and Magma of the Earth's Interior as a Water Supply.—Many geologists (Fisher, Tschermak, Reyer, Lane, and more recently Suess, Fairchild,* and Lapparent)† have expressed their conviction that the true source of volcanic water is to be found in the hydrated rocks and magma of the earth's interior, and there is much to support their view, even if it cannot be accepted to a complete exclusion of the consideration of the part which oceanic water may take as an assistant or aid. Unfortunately, the geologist is here again at the disadvantage of not being able to obtain hold of absolute facts, or of anything beyond plausible surmise. Neither negatively nor positively does the problem offer much fruit; nor, indeed, can it, until the equally obscure problem of oceanic penetration has in itself been resolved.

---

* Bull. Geol. Soc. America, January, 1904.

† L'Éruption de la Martinique, Revue des Questions Scientifiques, January 20, 1903.

INDEX

INDEX

PLATE I

Pelée in eruption, as seen from the graveyard of Marigot, about eleven miles in a direct line east by north of the crater, in the early morning of August 26, 1902. The crosses, head-stones, etc., are illumined by the horizontal rays of the rising sun. A marked effect upon the rising steam-ash cloud of the volcano produced by wind-action is seen in its abrupt bending over in the lower atmospheric zone. The upper reversed course is the result of the penetration of the higher column into a zone of contrary (antitrade?) winds—a condition that was not infrequent in the Pelée picture. The broad extent of the outflowing steam-ash cloud is beautifully shown here.

Expl. Heilprin.

I

PLATE II

Pelée, with its terminal tower or obelisk, as seen from the southern section of Saint-Pierre (looking north-northeast, and across an interval of about five miles). Total height of the mountain somewhat over 5000 feet, with the tower, which is moderately curved over in the direction of Saint-Pierre and shows a disrupted face turned to this side, constituting about one-sixth. The deeply-incised gully or rift descending from the base of the tower to the centre-middleground of the picture is the upper portion of the Rivière Blanche valley, into which much of the discharge product of the different eruptions was swept, and which marked the more general course of the descending *nuées ardentes* (black eruption clouds, of the French Scientific Commission). The dark heights on the right are Mont Parnasse, the summit and slopes of which were swept by the destroying blast of May 8. Photograph taken on a day exceptionally clear of volcanic storm.

III

PLATE III

The breaking clouds and vapors uncovering the giant tower, as seen from the south, a short distance below the summit of the mountain (June 13, 1903). The fingered, pinnacled, or serrated contour of the western side is well shown;—likewise, the indented apical summit. The lower slope on the left is the supporting cone or dome. Attention may be called to the long vertical rift appearing near the centre of the tower. Many of the greater breakages in the final destruction of the tower took place along lines of such fracture.

Photo. Heilprin.

III

PLATE IV

The great tower, as seen from the crater-rim in the afternoon of June 13, 1903, and looking north-northwest. The most instructive feature in this picture is the supporting cone or dome—the mass built up of lava and fragmental material which is implanted upon the floor of the basin of the Étang Sec and is the virtual new crateral-cone of the volcano. Steam and sulphur vapors are being puffed through its walls. The height of this dome, which in its later stages could be compared structurally with the cumuloid dome of Giorgios, in Santorin, surpassed by a hundred feet or more the crater-border immediately abreast of it. Two powerful ribs or ridges of continuous lava appear on the right-hand side, giving evidence of a true fluidal or semi-fluid condition.

Photo. Heilprin.

IV

PLATE V

The Tower of Pelée as seen from the old crater-rim, and exhibiting the side turned to Assier and Vivé (approx. east-northeast). Its cork-like extrusion from the cone or dome is impressively shown, and it at once suggests the probability of the mass being merely a pushed-out ancient core or plug. There is here no suggestion of a rapidly-solidifying (recent) viscous lava. The tower on this face is smooth and planed from top to base, unquestionably the result of attrition against the encasing wall of rock or "mold" which guarded the exit of the giant core. A portion of the surface is polished, and the greater part of it clearly exhibits longitudinal grooving and striation. The puffs of steam in the foreground are being blown through the mass of the cone. Somewhat more than 800 feet of the height of the tower are visible in the picture. June 13, 1903.

Photo. Heilprin.

V

PLATE VI

Views of and from the summit of the volcano (June 13, 1903). 1. Looking into the basin of the Étang Sec, whose restricted area is seen in the narrow circular or horse-shoe-shaped valley, brought to within some 300 feet of the summit of the mountain, which is interposed between the old crateral wall (left side) and the newly-constructed cone or dome (right-hand side of picture). Its width at the top may be fairly measured on the line of its depth. 2. The remains of the Morne de la Croix, the former highest point of the volcano. Its steep descent into the crater-basin (Étang Sec) is well marked. 3. The rim of the volcano (edge of the crater-wall of the basin of the Étang Sec). It looks down into the valley (*rainure* of the French Commission) seen in Fig. 1 and directly over to the central cone and its transfixing obelisk or tower. The steep plunge emphasizes the caldera structure. The declivity sloping off to the right, and largely covered with ejected bombs or boulders, is the slope descending into what was formerly the basin of the Lac des Palmistes.

VI

PLATE VII

Pelée and the ash-covered and deeply dissected valley of the Rivière Blanche, seen from about a half mile off shore on June 15, 1903. The front mud-wall is largely the mud-flow which on May 5, 1902, overwhelmed the Usine Guérin. The middle-ground is the more-recently fallen ash, almost white in color, and giving the appearance, especially in the darker hours, of newly-deposited snow. This ash deposit has a thickness in places of seemingly 200–300 feet and more. A light steam pennant is issuing from the absolute apex of Pelée's tower. On this day I observed this phenomenon almost uninterruptedly for upwards of an hour, and it was not difficult to determine that the pennant was truly issuing from the interior of the tower—which it, doubtless, traversed in rifts—and was not a normal mountain cloud of condensation ("mountain banner"). The absolute height of the volcano, from sea-level to its ultimate apex, was about 5000 feet, of which the tower made up almost exactly one-sixth.

VII

PLATE VII*a*

Pelée, after the destruction of its tower, as seen from Saint-Pierre (now being overgrown by new vegetation). The summit of the volcano clearly shows the basal wreck of the tower, which rises up as a low buttress above the crest of the mountain, and adjoins (on the left) the shortened pinnacle known as the Petit-Bonhomme. The right-hand slope of the volcano descends to Morne Rouge. Steeply descending from near the apex, with its further wall in shadow, is the gorge of the Rivière Blanche. Photograph taken in March, 1904, by J. Murray Jordan.

VIIᴀ

PLATE VIII

Pelée in full activity, in the afternoon of August 30, 1902; photograph taken about six hours before the cataclysm of the evening (nine o'clock), which destroyed or wrecked Morne Rouge and Ajoupa-Bouillon and partly devastated Morne Balai, Morne Capote, and the heights of Bourdon (Basse-Pointe). The vast swirling masses of steam, largely charged with ash, and sweeping out with swift velocity from the crateral basin as well as from the terminal cone, are well shown in their convoluted courses,—a picture of most impressive grandeur. At the position occupied by my party, on the upper slope of the volcano approaching the summit, little or no ash fell, the driving force of the volcano keeping the ash-umbrella floating at a dizzy height overhead. On our descent to the lower slopes, into the region of more natural calms, and at a greater radial distance from the crater, the ash fell over us in large quantities, mostly in a water-formed paste. It had the normal temperature of the air.

Expl. Heilprin.   Singley, Keystone View Co., Copyright, 1902.

VIII

PLATE IX

Pelée on the afternoon of August 30, 1902; photograph taken during a momentary sun-burst, a few hours before the death-dealing cataclysm of the same evening (see Plate XIII). The summit of the volcano, which is being neared by my party, is blotted out by the vast mountains of steam and ash that are being hurled out from the entire basin of the crater, and whose initial velocity on leaving the summit was timed to be from $1\frac{1}{2}$ to 3 miles per minute. I estimated the thickness of the steam column where it rose from the mountain to be not less than 1200–1400 feet, and it usually rose in vertical courses. The furious boiling of this storm-mass, rasping as it did the crateral walls, produced an indescribably terrifying—and one might truly say, appalling—roar, which can perhaps best be compared with the roaring of the ocean wind through the rigging and sheets of a fleet of vessels. A distinct vibration of the volcano was perceptible. I do not recall any electric flashes traversing this cloud-mass. Boulders and exploding bombs flew out in all directions.

Expl. Heilprin.

IX

PLATE X

Pelée in the early morning of August 31, 1902; as seen from our quarters at the Vivé estate (about five miles in a direct line east-northeast of the crater), and interesting as showing the forceful activity of the volcano *after* the cataclysm of the evening before (see Plates VIII. and IX.). The ash-cloud is being projected to a height of from three to four miles above the summit of the volcano, and is taking the more general north-northwest course out to sea. The great height attained by it is in the main the result of actual propulsion, and not of an ascensive force due to expansion and diminished weight. This same activity maintained itself seemingly unabated to and beyond the day of my departure from Martinique, a period of more than a week. It was for this period more particularly that, in a rough calculation, I estimated the daily discharge of ash to more than equal the *annual* discharge of sediment by the Mississippi River.

Singley, Keystone View Co., Copyright, 1902.

X

PLATE XI

Strongly-illumined steam-ash clouds projected in the afternoon of August 30, 1902, and beautifully exhibiting the convoluted structure, the rapidly unfolding whorls, which have been made part of the aspect of the so-called " cauliflower" volcanic cloud.

Expl. Heilprin.

XI

PLATE XII

Block of andesite, about four-fifths natural size, obtained from the eastern slope of Pelée, and with little doubt ejected during the eruption of August 30, 1902. It has the superficial resinous or semi-vitreous lustre of typical hypersthene-andesite ejected boulders, and the divisional cracks of bread-crust bombs. The weight is about that of normal compact rock, and the interior shows no truly cavernous or cellular structure. The block, like hundreds of others of this class lying on the summit and slopes of the volcano—some not larger than a marble and others 1-3 feet, and more, in diameter—is seemingly a part of the old stock of the volcano, which has been forced asunder, hard-rubbed by the escaping steam, and cracked by steam forcing itself into, and out from, it.

Photo. Heilprin.

XII

PLATE XIII

Morne Rouge after the destruction (photograph taken on September 7, 1902), with the church of Notre Dame de la Déliverance in the centre—one of very few buildings of the town that were left standing. It itself had part of its roof lifted and the right wall (not seen in the illustration) broken through. It was in passing from the presbytery of this church that Père Mary, the officiating parish priest, whose steady adherence to duty during the most trying Pelée days won for him the admiration of the world, was stricken by the blast of August 30. Among the débris of destruction that are scattered about the church-yard may be recognized, in the darker masses, the bodies of a few of the unfortunates who perished in the volcano's wrecking path.

Expl. Heilprin.

Singley, Keystone View Co., Copyright, 1902.

XIII

PLATE XIV

A portion of Morne Rouge, just off the traversing road, and adjoining the church of Notre Dame de la Déliverance—the wreck of the eruption of August 30, 1902. While the annihilation of the town was in a measure complete, leaving few houses intact, the force of the destroying blast was in this quarter inferior to that which shattered Saint-Pierre, as is evidenced by numerous frail objects—tree-trunks, posts, etc.—that had been left standing and that successfully defied the storm. On the other hand, the manner of the destruction appears to have been identical in the two cases. The loss of life resulting from the destruction of the town was probably 1000–1200, although the parish records place it at not less than 1500.

Photo. Heilprin.

XIV

PLATE XV

The village of Prêcheur, lying at the west base of Pelée, buried beneath its mantle of volcanic ash. This picture of extreme desolation is that of a winter landscape, contributed by the general whiteness or light-gray color of the fallen ash. Most of the destruction that appears was wrought by the weight of the deposited ash, which in places is many inches thick, and has broken through roofs and laid to flat measure the trees and shrubbery of the gardens. Devastating flood-waters have washed out parts of the town, and the church as it stands has had washed out from it one-half of its framework. On visiting Ajoupa-Bouillon and Morne Balai on the morning of August 31, 1902, immediately after the cataclysm of the night before, I found that many of the smaller houses had collapsed under their burden of ash, and large and small branches of trees had broken across in the manner of our northern trees when subjected to a heavy ice-coating. During the May eruptions many of the cocoanut palms were bowed down by ash miles away from the volcano.

XV

PLATE XVI

Looking up the valley of the Rivière Blanche, and into the V-shaped cleft through which the destroying blast is thought to have issued from the crater basin on May 8 (in the furthest centre). The upper moiety of the volcano is buried in cloud and vapor, but the base of the cone appears in the cleft. The floor of the valley is filled with volcanic débris to a depth in places of not less than 200–250 feet—possibly considerably more— and this has in the main accumulated as the result of the August 30 (1902) and later eruptions. Most of the giant blocks of rock that appear scattered over the valley, measuring anywhere from five and ten to 40–50 feet in length, were ejected during the outburst of August 30, as I had occasion to ascertain by direct observation a few days after that event. There would seem to be no way of positively fixing the method of deposition of these rock-masses—of ascertaining if they were simply shot or rolled down the mountain slope, or flung through the air to the positions which they now occupy. Some of the blocks are completely fractured across, which would seem to indicate a fall. The members of the Lacroix mission examining one of the largest of the ejected blocks.

XVI

PLATE XVII

The gorge of the Falaise in its middle course, where in a cut with vertical sides it has narrowed to a few yards' width. The bounding walls, which are an earlier volcanic (agglomeritic) accumulation, readjusted by (oceanic?) water, are the supply source of much of the boulder-débris which encumbers the lower Falaise, and which at the junction with the Capote was deposited by the volcanic flood-waters to a height of 15–20 feet. The tempestuous force of these minor streams coursing down or away from the volcano's foot was extraordinary, acres of boulders having been hurled along almost as if they were floating blocks of wood. Many of these were 5–8 feet in length. some considerably more, and one (andesite) measured by me on August 28, 1902, was 16 feet in length, nearly 10 feet in width, and 6–8 feet thick.

Photo. Heilprin.

XVII

PLATE XVIII

Fragments of manuscript, dealing with volcanic phenomena as part of a geological lesson, recovered from the débris at Saint-Pierre. The darker portion of the photo-engraving gives the contour of the paper, which is perforated by holes and gashes. The condition of Pompeii at the time of its destruction is referred to in the first line of the upper figure.

...ont été conservées. Pompéi était alors situé[e] ↑ le
bord de la mer ; aujourd'hui elle en est distante de
600 m, car petit golfe comblé par les cendres.
Le moteur de la lave est la vapeur. Il faut
assimiler la lave à une éponge à l'état sphéroï-
dal [sursaturée] d'eau surchauffée, et on voit une
...lée émettre de l... ... jusqu'à solidific...
...fora élastique de la vapeur
...roulements de la vapeur, alo...
...coulée détermin... éruption
...la lave arrive seule
...ule (rare).
...peu dans le cratère
...large cavités, puisqu...
...de plus la
—3 D.

répand sur le sol. La lave est très gênée dans ses
mouvements. La surface se fige sans cesse. Il faut
qu'elle brise l'extrémité du sac pour pouvoir avancer.
Le rayonnement de la coulée est nul, bien que la
pellicule solide soit mince. C'est aussi là la raison
de la lenteur du refroidissement ; il faut 4 ou 5 ans pour
[1] coulée 3 ou 4 m d'épaisseur, si 8 m (Etna) il faut
... Par les fentes de la coulée, on voit
...l... vapeur d'eau.
...duits. Ce sont les fumerolles
du volcan.
...lave est encore liquide
...fumerolles riches ;

Photo. Heilprin.

XVIII

PLATE XIX

Additional fragments of manuscript, in two leaves, recovered from the débris at Saint-Pierre in June, 1903, and interesting in their reference to volcanic phenomena. Manifestly it is a student's copy of a lesson in geology, and possibly represents part of a course in the last days of the Lycée or the Communal College. We found these leaves beneath boulder-masses near the centre of the town; they were turned yellow or brown and burned only on some of the edges.—Photographic copy.

Le jour, le bruit est sensible. Au Cotopaxi (Pérou)
le fait se produit toujours; le bruit retentit à 900
kilomètres du volcan; près de la ville le Honda
-velle grande) et cependant il n'y a rien; ...
internes...

On a longtemps ignoré l'origine des cendres; ...

XIX

PLATE XX

Statuette, in bronze, about four-fifths full size, recovered from Saint-Pierre, and obtained by purchase in Fort-de-France. The "pocked" and pitted appearance of the surface is due to the adhesion—an "inwelding"—of particles of volcanic ash to the substance of the object, the result of heat action. The rearing horse is part of the group of the famous "Horse Tamer."

Photo. Heilprin.

XX

PLATE XXI

Water-bottle or caraffe from Saint-Pierre, about four-fifths full size, and interesting as showing marked deformations of its substance without breakage. There are no indications of glass-flow, and the only apparent change that the glass has undergone is an acquired murkiness. The substance had evidently yielded to pressure-impacts at a time when it was subjected to and softened by great heat. This condition, which is also represented in the wine-glasses figured in the following plate (Plate XXII.), sufficiently explains the similar condition of objects found at Pompeii, and does away with the necessity of assuming that the deformation was the result of a slow and steadily progressing molecular change whose workings had extended through centuries (!)

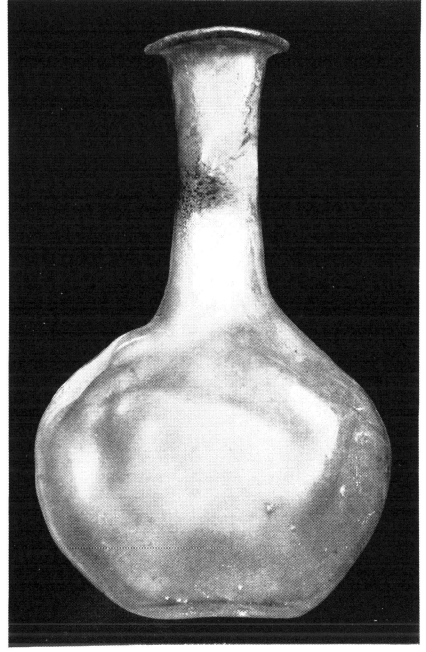

Expl. Heilprin.

XXI

PLATE XXII

Deformed wine-glasses from Saint-Pierre. See explanation of Plate XXI.

XXII

Printed in the United States
By Bookmasters